Geology of the Bitterroot and Clearwater Mountains of Idaho and Montana

by United States Geological Survey

with an introduction by Kerby Jackson

Introduction

It has been years since the United States Geological Survey released this important publication. First released in 1904, this important volume has now been out of print for this days and has been unavailable to the mining community since those days, with the exception of expensive original collector's copies and poorly produced digital editions.

It has often been said that "*gold is where you find it*", but even beginning prospectors understand that their chances for finding something of value in the earth or in the streams of the Golden West are dramatically increased by going back to those places where gold and other minerals were once mined by our forerunners. Despite this, much of the contemporary information on local mining history that is currently available is mostly a result of mere local folklore and persistent rumors of major strikes, the details and facts of which, have long been distorted. Long gone are the old timers and with them, the days of first hand knowledge of the mines of the area and how they operated. Also long gone are most of their notes, their assay reports, their mine maps and personal scrapbooks, along with most of the surveys and reports that were performed for them by private and government geologists. Even published books such as this one are often retired to the local landfill or backyard burn pile by the descendents of those old timers and disappear at an alarming rate. Despite the fact that we live in the so-called "Information Age" where information is supposedly only the push of a button on a keyboard away, true insight into mining properties remains illusive and hard to come by, even to those of us who seek out this sort of information as if our lives depend upon it. Without this type of information readily available to the average independent miner, there is little hope that our metal mining industry will ever recover.

This important volume and others like it, are being presented in their entirety again, in the hope that the average prospector will no longer stumble through the overgrown hills and the tailing strewn creeks without being well informed enough to have a chance to succeed at his ventures.

Kerby Jackson
Josephine County, Oregon
October 2015

CONTENTS.

4 CONTENTS.

ILLUSTRATIONS.

LETTER OF TRANSMITTAL.

DEPARTMENT OF THE INTERIOR,
UNITED STATES GEOLOGICAL SURVEY,
Washington, D. C., June 3, 1903.

SIR: I have the honor to transmit herewith the manuscript of a report entitled "A Geological Reconnaissance across the Bitterroot Range and Clearwater Mountains in Montana and Idaho, by Mr. Waldemar Lindgren." The report contains the results of a rapid reconnaissance made in August, September, and October, 1899, covering an area of 12,000 square miles in Idaho, Montana, and Oregon. The region is one concerning the geology of which little information is available, and this report, while merely a reconnaissance, should be of material assistance in the development of the economic resources of the region.

Very respectfully,

C. W. HAYES,
Geologist in Charge of Geology.

Hon. CHARLES D. WALCOTT,
Director United States Geological Survey.

8

A GEOLOGICAL RECONNAISSANCE ACROSS THE BITTERROOT RANGE AND THE CLEARWATER MOUNTAINS IN MONTANA AND IDAHO.

By WALDEMAR LINDGREN.

INTRODUCTION.

SITUATION.

This report describes, in a preliminary way, a belt of country extending westward from the Bitterroot Valley, across the dividing range and the rugged mountains of the Clearwater system, down to the fertile plateaus which border the canyon of Snake River. It thus presents a reconnaissance section from western Montana across northern Idaho, and deals chiefly with areas about which, thus far, little geological information has been available.

In spite of the forbidding aspect of these mountains, the earliest explorers of the Cordilleran region, Lewis and Clarke, led their expedition across them, choosing the Lolo trail from the Bitterroot Valley to the present site of Lewiston, where they wintered. This was in 1805. For many decades after this the wild solitudes remained the exclusive hunting ground of the Nez Perce Indians; but in 1860 the gold fields of Elk City and Oro Fino were discovered, and in that year began the gradual settlement of the Bitterroot Valley on the east and the lower Clearwater country on the west. An army of gold seekers invaded the Clearwater Mountains, but, failing to find gold placers outside of the old camps, soon abandoned the search. The two Indian trails connecting Idaho with Montana, the Lolo and the Nez Perce, which were much used at first, gradually became less frequented. The Indians and the few trappers who wintered in the snowy forests were the only occupants of these mountains, which, on account of their isolation, formed one of the finest game preserves of the West, where the moose, the elk, and the bear roamed in great numbers.

In 1877 the lower Clearwater country was the scene of the Nez Perce Indian war, and the march of the United States troops across the Lolo trail after the Indians forms one of the memorable episodes of this unfortunate conflict. In 1896

9

there was established the present United States forest reserve, which embraces the larger part of the Clearwater drainage. During the last few years the reserve has been patrolled by Government rangers in order to prevent forest fires, which in the past have caused such destruction within this area, and in order to check the indiscriminate slaughter of wild game which has gone on during the last ten or fifteen years. The Bitterroot Valley has gradually been settled, and to-day its orchards and fields form one of the most attractive parts of Montana. In 1897 the Nez Perce Reservation, which had long been a barrier to the settlement of the lower Clearwater, was thrown open, and the high prairies between the Clearwater and the Snake have been transformed into a rich and important farming country.

MAPS AND LITERATURE.

Until a recent date the topography of the Clearwater Mountains was only imperfectly known. The first approximately reliable data are contained in the map of the Department of the Columbia issued in 1881 by the Engineer Corps, United States Army, which was compiled by Lieut. T. W. Symons. During 1897 Mr. J. B. Lippincott made, for the United States Geological Survey, a reconnaissance map of the newly created forest reserve, on the scale of 6 miles to the inch. This excellent work is the chief basis of the maps published since then and is also used in this paper, somewhat reduced in scale. Many of the photographs in this report were taken by Mr. Lippincott in 1897. In 1897–98 the Hamilton quadrangle (Montana and Idaho) was mapped by Messrs. S. S. Gannett, E. C. Barnard, and A. B. Searle for the United States Geological Survey. The map of this quadrangle includes the crest of the central part of the Bitterroot Range; it is published on the scale of 1:125000, or about 2 miles to 1 inch, and it has 100-foot contour intervals. These, together with the cadastral surveys of the United States General Land Office (General Land Office, Department of the Interior, map of Idaho, 1898) of the eastern and western margins of the area involved, form the basis of the maps accompanying this report. Pl. I, which shows the region covered by this reconnaissance, is compiled from the Geological Survey map by Lippincott and the Land Office map of Idaho. The contours west of the western boundary of the forest reserve were sketched by myself from numerous aneroid observations.

The reports dealing with the whole or parts of this region are as follows:

RUSSELL, J. C., A reconnaissance in southeastern Washington: Water-Sup. and Irr. Paper No. 4, U. S. Geol. Survey, 1897.
    Contains description of the canyon of Snake River below Lewiston.
GOODE, R. U., The Bitterroot Forest Reserve: Nat. Geog. Mag., vol. 9, 1898, pp. 387–400, No. 9.
    Contains general description of reserve and definition of mountain systems.

LEIBERG, J. B., The Bitterroot Forest Reserve: Nineteenth Ann. Rept. U. S. Geol. Survey, pt. 5, 1899, pp. 253–282.

Describes topography and forests of the Clearwater Mountains.

LINDGREN, WALDEMAR, The gold and silver veins of Silver City, De Lamar, and other mining districts in Idaho: Twentieth Ann. Rept. U. S. Geol. Survey, pt. 3, 1900, pp. 67–256.

A reconnaissance south of the Clearwater Mountains and partly connecting with the area described in the present report. Describes also the gold-mining camp of Florence and the canyon of Salmon River west of it.

RUSSELL, I. C., Geology and water resources of Nez Perce County, Idaho: Water-Sup. and Irr. Papers Nos. 53 and 54, U. S. Geol. Survey, 1901.

Describes the lower Clearwater region, chiefly occupied by the Columbia River lava. Also contains accounts of underlying formations and of the canyons of Salmon and Snake rivers.

WHITTLE, C. L., The Buffalo Hump mining district: Eng. and Min. Jour., vol. 68, 1899, pp. 215, 216.

Contains a short description of this mining camp.

FINLAY, J. R., The mining industry of the Cœur d'Alene, Idaho: Trans. Am. Inst. Min. Eng., February and May, 1902.

Many scattered notes regarding mining operations and production are found in the early reports to the United States Government by J. Ross Browne and R. W. Raymond, as well as in subsequent reports by the Bureau of the Mint on the production of gold and silver.

FIELD WORK.

The area embraced in the present reconnaissance contains about 12,000 square miles, of which 6,500 square miles are included in the Bitterroot Forest Reserve. Beginning at the confluence of Bitterroot River and Lolo Fork, the boundary of this area extends south to the head of the South Fork of the Bitterroot River; then it follows the divide between the Salmon and the Clearwater to Buffalo Hump; then it runs northwestward to Mount Idaho, Grangeville, and the lower canyon of the Salmon, then to Lewiston, and from there eastward to Pierce; then it follows the divide between the Lochsa Fork and the North Fork of the Clearwater to Lolo Pass, and down the Lolo Fork to its mouth. (Pl. I and fig. 1, p. 12.)

The field work occupied the time between July 22 and October 21, 1899. During this time Mr. H. R. Johnson, of Stanford University, rendered efficient geological aid, and during my temporary absence, from September 22 to October 8, had charge of the party examining the western part of the Lolo trail and the Crags. From August 1 to September 11 I had also the valuable assistance of Mr. G. W. Stose, to whom especially many observations on the upper Bitterroot Valley are due. The general route led first north from Hamilton along the Bitterroot Range to some miles beyond Lolo Pass, thence back to Hamilton and across the range over the Lost Horse trail. Returning, the southern part of the range was examined, and from Darby the route led up the Bitterroot River to its head; thence along the divide between the Salmon and the Clearwater to Elk

City and Buffalo Hump; thence across the high prairie south of Lewiston, up the main fork of the Clearwater to the Crags; finally, to Pierce and up to Rocky Ridge on the Lolo trail. Many parts of the area were not visited and the whole examination partakes of the character of a rapid reconnaissance. Fortunately,

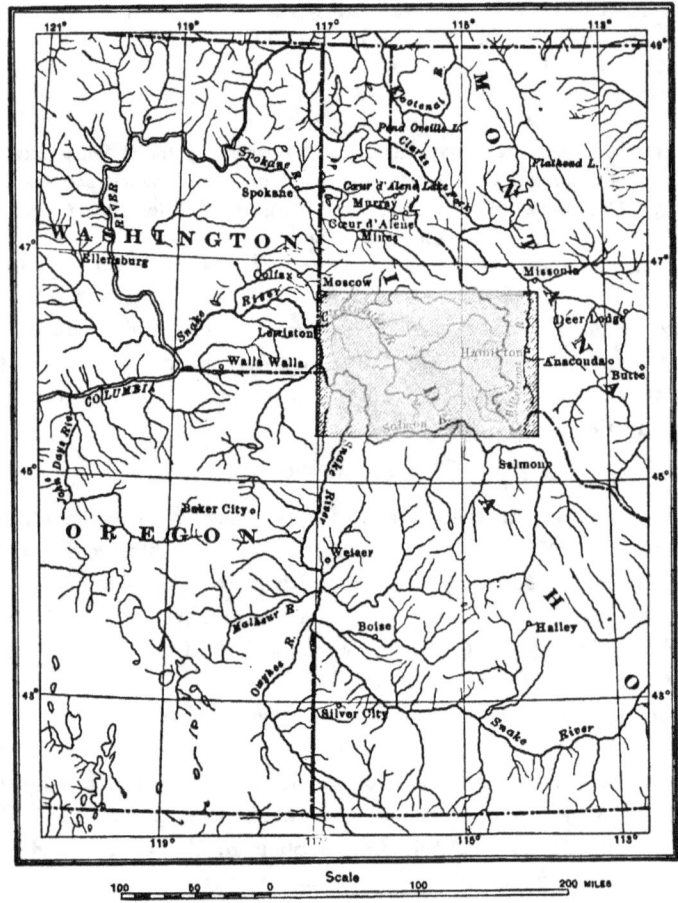

FIG. 1.—Index map showing extent of reconnaissance.

the route was laid out in a manner to obtain, in spite of this, a fairly accurate conception of the general geological structure of the whole region.

TOPOGRAPHY.

The nomenclature of the main ranges in these parts of Idaho and Montana is not yet settled, and it will, therefore, first be necessary to define the mountain systems involved. The whole area lies in the watershed of the Columbia River.

As is well known, the main divide of the Rocky Mountains is followed by the boundary line between Montana and Idaho from the Yellowstone Park westward to Gibbons Pass, near the southeastern corner of this reconnaissance area; thence the main divide bends sharply northeast, while for over 200 miles northward the boundary follows the divide between the tributaries of Snake River on the west and Clark Fork of the Columbia on the east, the principal affluent of the latter being the Bitterroot River.

It is proposed to give the name of the Coeur d'Alene Mountains to the broad complex extending 100 miles from Pend Oreille Lake southeasterly to the head of the North Fork of the Clearwater, a short distance north of Lolo Pass. This range, which reaches elevations of only 7,000 or 7,500 feet, is drained westward by the Coeur d'Alene, St. Joseph, and St. Marys rivers, all emptying into the Coeur d'Alene Lake, and thence into the Columbia by way of Spokane River. At the western foot of this range extend the high plateaus of the Palouse country. The eastern slope of the Coeur d'Alene Mountains is only from 10 to 25 miles wide, and the creeks drain directly into Missoula River or the Clark Fork.

From Lolo Fork to the head of the West Fork of the Bitterroot River (also called Nez Perce Fork) extends the Bitterroot Range. It is extremely well defined, topographically and geologically, is about 60 miles long, and trends exactly north and south. In many places it attains elevations of over 9,000 feet, and near the southern end culminates in Trapper Peak (elevation, 10,175 feet).

Eastward the Bitterroot Range descends steeply toward the valley of the same name. Westward its slopes soon merge into the Clearwater Mountains.

By the name Clearwater Mountains it is proposed to designate that broad, mountainous country comprised within the drainage of the Clearwater River. The Clearwater Mountains occupy a rough square with a side of about 70 miles, and at their westward margin descend rapidly to the high plateaus of the lower Salmon and Snake rivers. The highest peaks rarely exceed 8,000 feet in elevation.

The Clearwater Mountains are continued southward in the Salmon River Mountains. These have the same general characteristics and occupy a large portion of central Idaho. The Salmon River is supposed to be the dividing line between these two regions.

The divisions here outlined are on the whole similar to those used by the United States Engineers in their map of the Department of the Columbia of 1881, but differ from those employed on the map of Idaho of the General Land Office, on which the Bitterroot Mountains cover the area between Mullans Pass in the Coeur d'Alene Mountains and Nez Perce Pass.

R. U. Goode (see list of literature, p. 10) gives the general name of Bitterroot Mountains to the whole uplift from near Pend Oreille Lake to the pass at the head of Bitterroot River, near the North Fork of the Salmon, and makes the Coeur d'Alene and Clearwater mountains subdivisions of this general term. On account of the strongly pronounced individuality of that part of the range between the Lolo and Nez Perce passes, which unquestionably has always been designated by the name of the "Bitterroot Range," I do not like to accept so wide a definition.

The Clearwater Mountains thus form a part of the great mountain region of central Idaho, a vast uplift extending from the Snake River Valley on the south well up toward the British possessions on the north. Like the Salmon River Mountains on the south and the Coeur d'Alene Mountains on the north, they are deeply dissected by a canyon system of rarely equaled extent and depth. The principal ridges between the streams, generally broad and flat, but sometimes also dissected into sharper crests and peaks, attain throughout about the same elevation, averaging perhaps 7,000 feet, and varying between 8,000 and 6,000 feet. Their combined crest lines would form an undulating plain differing little in elevation in the various parts of the Clearwater Mountains; in other words, sloping very little, if any, in any given direction. From their westerly margin the mountains slope rapidly to the lava plateau, which has an elevation of 3,000 to 3,500 feet. Along the Salmon River the high mountain plateau extends farther westward, and its last ramparts overlook the great bend of that river, rising 6,500 feet above its water line. The plateau character of the combined ridge lines is clearly seen in figs. 4 and 6 (pp. 60, 69). We must regard this surface as the result of erosion. The country was worn down to a comparatively gentle topographic feature, then uplifted and deeply dissected by canyons. Foremost of these is the great canyon of the Salmon, which, like a mighty V-shaped trench, is cut to a depth of from 4,000 to 5,000 feet practically across the whole width of Idaho, 100 miles in a straight line, and far more along the course of the river. It interposes a formidable obstacle to intercourse between the southern and northern parts of the State. Only one wagon road crosses it, and that can be traversed by only lightly loaded vehicles. Within a few miles of Snake River the Salmon bends suddenly northward and parallels the Snake for a distance of 40 miles, and then finally joins it, after making a wide curve convex toward the north. Until a point some distance below the great bend is reached the canyon is entirely cut in pre-Tertiary rocks and antedates the flows of the Columbia River lava. No bottom lands worthy of the name line the main part of the canyon, and many parts of the gorge are entirely inaccessible. Disaster has frequently overtaken those who have tried to navigate its wild rapids. The river is still largely eroding its bed. Below Freedom its course was changed by the Tertiary lavas.

The lower portion of the Clearwater River in the Columbia River lava closely parallels the Salmon, and its course is undoubtedly due to the same determining cause—that is, to the obstructing Tertiary flows of basalt. The Clearwater drainage system is much more branching, and though the trenches of the principal forks, the Lochsa and the Selway, are fully 4,000 feet deep and often inaccessible along the stream beds, they lack the grandeur of the single canyon of the Salmon. As seen from the map (Pl. I), the Clearwater splits in a number of branches, and, in fact, drains practically the whole area discussed in this paper. When the river enters the plateau of Columbia River lava the canyons are much less deep, but are much more sharply incised than the flaring trenches of the Clearwater Mountains.

Between the Lolo and Nez Perce passes, very close to the Montana boundary line, the mountains increase rapidly in elevation and form the narrow and sharply defined Bitterroot Range, which eastward soon slopes down to the equally pronounced depression of Bitterroot Valley.

The Bitterroot Range can best be likened to a long and narrow block raised 2,000 feet above the general elevation of the old Clearwater Plateau. North and south of it fragments of this plateau seem to continue for a still undetermined distance into Montana.

The Bitterroot River heads on the ridge north of the Salmon River Canyon. In its upper course, as far north as the southern end of the Bitterroot Range, the river and its branches have evidently dissected the same old Clearwater Plateau. It soon, however, enters the wide and level Bitterroot Valley and continues northward in this for over 60 miles. North of the mouth of Lolo Fork the bordering ridges again rise to the general level of the Clearwater Plateau. Gently sloping ridges face the Bitterroot Valley on the east, while on the west rises the high block of the Bitterroot Mountains, sloping down to the valley by a long, even declivity of 18°, and indicating by its geologic and topographic features its important tectonic character as a fault plane.

DESCRIPTIVE GEOLOGY.

GENERAL FEATURES.

SEDIMENTARY FORMATIONS.

In a broad way the distribution of the pre-Tertiary sedimentary formations in the region under consideration is as follows: Practically the entire area of the Bitterroot and Clearwater mountains is occupied by granite with some gneiss. West of the Clearwater River, and only imperfectly exposed below the lava, is an extensive sedimentary area adjoining this granite; smaller sedimentary areas are exposed on Lolo Fork and on the head of the South Fork of Bitterroot River. In no place have well-defined fossils been found, but there is some foundation for the belief that the two last-named areas on the east side are very old, possibly pre-Cambrian, while the western area probably includes Triassic, Carboniferous, and possibly still older sediments.

The sedimentary series of the Lolo Fork, in the northeast corner of the area, consists of moderately metamorphosed quartzites, limestone, and banded slates, some of them of a purple color. This apparently conformable but evidently folded, over-turned, and repeated series has a northwesterly or west-northwesterly strike and a prevailing northeasterly dip of 30° to 45°, and borders south and west against later granite with intrusive contact. The chief interest of this series lies in the fact that it is probably the southward end of the similar very extensive area which in the Coeur d'Alene section, 75 miles farther north, occupies the whole width of the Coeur d'Alene Mountains from the lake of the same name up to Mullan Pass, and thence down to the Clark Fork of the Columbia by way of St. Regis de Borgia River. Throughout, this Coeur d'Alene series has a west-northwest or northwesterly strike and usually a moderate northeasterly dip. No fossils have been found in it, and my belief is that it should be coordinated with the thick pre-Cambrian series oᵢ central Montana.

Small exposures of the same or a similar series are noted along the foothills of the Bitterroot Valley and again on the Skalkaho and Weeping Child Creek, where they also border intrusive granite.

A series of quartzites or quartzitic schists and slates occupies a considerable area at the very head of the South Fork of the Bitterroot and extends at least as far as Gibbonsville, Idaho, on the east. They appear to be thrown into flat folds dipping east or west at moderate angles. They are probably intruded by granite and certainly by granite-porphyry on the north, and border against older gneisses on the south.

At the western foot of the Clearwater Mountains the most prominent sedimentary terrane is composed of black slates, limestones, quartzitic slates, and associated partly schistose and probably effusive greenstones which are found on the South Fork of the Clearwater and which extend only a short distance north of Harpsters. This series has a persistent northeasterly strike and steep dip, and evidently forms the northern continuation of beds of the same strike exposed at intervals below the Columbia River basalt on the lower Salmon, in the Seven Devils, and, over 100 miles distant, near Huntington. Triassic fossils have been found in the Seven Devils,[a] and a large part of this long area of sediments is believed to be of Triassic age. Still farther west sedimentary rocks of unknown age, but probably Paleozoic or Mesozoic, are exposed near the junction of the Salmon and Snake rivers, in the Cottonwood Buttes, and in Clearwater Canyon

---

[a] Twenty-second Ann. Rept. U. S. Geol. Survey, pt. 3, 1902, p. 581.

between Contact railroad station and the mouth of the North Fork. They consist of quartzites, sandstones, conglomerates, limestones, and shales, as well as large masses of associated old and partly altered and schistose lavas (greenstones). The structure of this series is little known except that the dips are generally steep and the strike has a general northerly direction.

At Oro Fino, North Fork, and Pierce are gneisses, mica-schists, and occasional crystalline limestones bordering the intrusive granite of the Clearwater Mountains. These are surely partly at least of sedimentary origin, but their age is unknown.

GRANITE.[a]

Extent.—Granitic rocks prevail in the Bitterroot Range and in the Clearwater Mountains, and form a central mass of vast extent, bounded in the four corners of the region covered by this reconnaissance by smaller areas of different sedimentary series. To the north of this region the extent of the granite is not well known. But as the granite is absent in the Coeur d'Alene section (p. 81) it is probable that the main area does not continue far north of Lolo Ridge except as detached masses. Southward this granite continues through all of central Idaho as a broad belt, and finally disappears below the sediments of Snake River Valley. It does not reach Snake River at any place between Huntington and Lewiston. It forms on the whole an elongated area 300 miles from north to south and 50 to 100 miles from east to west, constituting one of the largest granitic batholiths of this continent.

Within the margins of this reconnaissance it occupies the whole of the Bitter-root Range except its extreme northern end. Along the foot of this range it acquires gneissoid structure, owing to orographic forces. It even appears in the eastern foothills of the Bitterroot Valley, and is there also slightly gneissoid in structure. Westward it occupies the whole center of the Clearwater Mountains from Lolo Ridge down to Salmon River, but along the latter contains many included areas of an older gneiss. Between Oro Fino and Stuart, on the Clearwater, a projecting arm of the granite extends far westward underneath the lava to the southeastern part of the Cottonwood Buttes, and possibly even to the Salmon River Canyon west of Grangeville.

Character.—On the whole, this extensive area of granite shows great constancy in its petrographic character. It is a normal granular rock, sometimes roughly porphyritic by the development of large orthoclase crystals up to 3 cm. in diameter. The color is almost always light gray, the outcrops assuming a yellowish-gray color, which in glaciated districts changes to a brilliant white or light-gray tone. Biotite is always present in small foils, and over large areas muscovite also enters into the

---

[a] The coarser granular, massive, igneous rocks occurring in this region consist chiefly of quartz-monzonites with frequent facies of true granite with prevailing alkali feldspar. The two kinds of rocks can not easily be distinguished in the field, and it has therefore been found necessary to speak of them collectively as "granite." The distinction between them is emphasized only in detailed petrographic descriptions.

composition; quartz is abundant in medium-size grains, while the feldspars are represented by both orthoclase and oligoclase, the latter usually in large quantity. Perthite is also frequently encountered, and more rarely microcline. The rock contains far too much oligoclase to be classed as a normal granite and should be rather characterized as a quartz-monzonite. Modifications more closely allied to granodiorite, diorite, and granite occur in subordinate quantity.

The granite is typically developed near the head of Mill Creek, Bitterroot Range, where it is a light-gray, medium-grained rock, with small foils of biotite and a little muscovite. A few larger crystals of orthoclase reach one-half inch in length. Under the microscope the rock shows much quartz, a little normal orthoclase, and many large grains of microperthite. An acidic oligoclase with very narrow striations is very abundant. Biotite and muscovite occur in scattered straight foils. Few accessories except zircon and apatite were noted, though titanite occurs abundantly in basic concretions in the same granite. The structure is typically granitic; the oligoclase is in part idiomorphic and sometimes included in the perthite. An analysis of this rock is given in the table below, together with such analyses of rocks as are available from the southern end of this great granitic batholith.

Analyses of granitic rocks from Idaho.

	I.	II.	III.	IV.	V.
SiO$_2$	72.07	69.56	68.42	65.23	57.78
Al$_2$O$_3$	15.51	15.29	15.01	16.94	16.28
Fe$_2$O$_3$	.31	.86	.97	1.60	1.02
FeO	1.01	2.06	1.93	1.91	4.92
MgO	.35	.69	1.21	1.31	4.60
CaO	1.93	2.81	2.60	3.85	6.65
Na$_2$O	4.02	3.97	3.22	3.57	3.25
K$_2$O	4.09	3.36	4.25	3.02	2.22
H$_2$O below 110° C	.03	} a.86 {	.54	.18	.34
H$_2$O above 110° C	.30		.73	.88	.92
TiO$_2$	.16	.55	.50	.66	1.07
CO$_2$	None.	....	.20	.25	.15
P$_2$O$_5$	.11	.16	.13	.19	.30
S	....	....	.02	....	.02
NiO and CoO	....	....	None.	....	.02
MnO	Trace.	....	.06	Trace.	.15
BaO	....	....	.12	.19	.12
SrO	....	....	.03	....	.07
Li$_2$O	....	....	Trace.	....	Trace.
	99.89	100.17	99.94	99.78	99.88

a Loss on ignition.

I. Quartz-monzonite. Near head of Mill Creek, Bitterroot Range, Montana. W. F. Hillebrand, analyst.

II. Quartz-monzonite, No. 46, Boise collection. Shafer Butte, Boise County, Idaho. George Steiger, analyst. Twentieth Ann. Rept. U. S. Geol. Survey, pt. 3, 1900, p. 82.

III. Quartz-monzonite, No. 25, Hailey collection. Democrat mine, near Hailey, Blaine County, Idaho. W. F. Hillebrand, analyst. Twentieth Ann. Rept. U. S. Geol. Survey, pt. 3, 1900, p. 82.

IV. Granodiorite, No. 79, Boise collection. Silver Wreath tunnel, Willow Creek mining district, Boise County, Idaho. George Steiger, analyst. Twentieth Ann. Rept. U. S. Geol. Survey, pt. 3, 1900, p. 82.

V. Diorite, No. 32, Hailey collection. Crœsus mine, near Hailey, Blaine County, Idaho. W. F. Hillebrand, analyst. Twentieth Ann. Rept. U. S. Geol. Survey, pt. 3, p. 82.

The approximately calculated composition of the rocks is as follows:

Calculated composition of rocks from Idaho.

	I.	II.	III.	IV.	V.
Quartz	28.73	28.04	29.21	25.00	8.45
Orthoclase molecule	20.04	15.84	18.07	11.21	7.57
Albite molecule	33.98	33.54	27.19	30.25	26.20
Anorthite molecule	8.36	11.15	9.53	13.88	20.45
Biotite [a]	[b] 8.03	7.55	12.36	15.99	[c] 32.55
Apatite	.27	.37	.47	.44	.69
Titanite	.29	1.13	.88	1.65	1.29
Magnetite	.17	.93	.31	.61	1.48
Calcite			.45	.57	.33
Pyrite			.03		.03
Hygroscopic water	.03	.36	.54	.18	.34
Excess combined water		.27	.32		.51
Total	99.90	99.18	99.36	99.78	99.89

[a] Molecular ratio assumed: MgO: SiO$_2$: TiO$_2$: Al$_2$O$_3$: Fe$_2$O$_3$: FeO: CaO: K$_2$O: H$_2$O=1: 2.44: .06: .75: .14: .88: .04: .42: .76.

[b] Biotite, 5.03; muscovite, 3.00. Molecular ratio of biotite in this analysis figures out to MgO: SiO$_2$: Al$_2$O$_3$: Fe$_2$O$_3$: FeO: CaO: K$_2$O: H$_2$O: TiO$_2$=1: 3.64: 1.11: .14: 1.52: .04: .42: 1.04: .06

[c] Biotite, diallage, hypersthene, hornblende.

The analysis of the rock from the Bitterroot Range shows so great similarity to II and III, both of which are representative rocks, that no special calculation seems necessary. It contains the same amount of quartz and a more sodic oligoclase, while the potash feldspar is slightly increased, a little muscovite being present in I. Analyses IV and V represent local basic modifications. The five analyses form a series with gradually lowered SiO$_2$, constant Al, and increasing Fe, Mg, Ca. The first three rocks should doubtless be considered as quartz-monzonites, though I distinctly approaches the granites.

In the Bitterroot Mountains streaks of basic magmatic segregations often appear, characterized by biotite and hornblende, quartz, orthoclase, oligoclase, and andesine, together with much titanite. Almost normal granites with prevailing orthoclase were observed west of Little Salmon Meadows and on the head of Crooked River. Dioritic modifications were observed near the schist contact at Pierce, and in the Clearwater Canyon near Greers Ferry. At the latter place a very fine diorite-pegmatite, consisting of hornblende and a plagioclase of medium acidity, is exposed in a railroad cut near the river.

Age.—The intrusive character of the granitic batholith is apparent at all places where it adjoins adjacent formations. The petrographic character is, on the whole, so constant that it seems probable that the whole area is of the same age, and a geological unit. The age can not be determined with certainty on account of the absence of fossils in the surrounding formations. In the southern part of the batholith, near Hailey, on Wood River, it has been shown that the intrusion is certainly post-Carboniferous.[a] As it has been shown (p. 16) that the sedimentary series on the South Fork of the Clearwater, near Harpster, is very probably Triassic, a post-Triassic age may, with the same degree of certainty, be attributed to the great granitic batholith.

Dikes in granite.—On the whole, dikes are not common in the principal granite area embraced in this report, even pegmatites and aplites being rare. In the eastern foothills of the Bitterroot Valley, however, and along the South Fork of Bitterroot River dikes of granite-porphyry and rhyolite are very abundant in the granite. The shattered gneiss areas of Horse Creek and Salmon Mountain, as well as the more compact areas of Elk City, contain also an abundance of pegmatitic, granitic, and dioritic dikes. The gneiss is sometimes injected by a great number of narrow pegmatite dikelets which have suffered a crumpling at the same time as the surrounding foliated rocks.

GNEISS.[b]

In a prevailing terrane of intrusive granite the Bitterroot and the Clearwater mountains contain numerous areas of gneissoid rocks of very differing ages. Some are later than the granite, of which they indeed, are only a modification, caused by peculiarly applied stresses. If the conclusion as to the age of the granite is correct, these gneisses are post-Triassic. Others are of much greater age, being probably older than the associated sediments, which are themselves older than the granite. In these sediments no fossils have been found, and their true horizon is thus doubtful. Very likely, however, parts or all of this second series of gneisses may be of pre-Cambrian age.

The first and most extensive gneiss area is that which follows the well-marked eastern slope of the Bitterroots, and which is clearly nothing but a sheet

---

a Twentieth Ann. Rept. U. S. Geol. Survey, pt. 3, 1900, p. 79.
b The word gneiss is here used without limitations as to age and origin.

of pressed and deformed granite. It is usually a light-gray, granular, and schistose rock, and consists of lenticular squeezed aggregates of oligoclase and quartz and deformed orthoclase crystals, separated by somewhat curved aggregates of new-formed biotite and muscovite. The primary constituents are the same as those of the granite. All transitions toward massive granite may be seen. Excessive compression results in finer-grained and darker micaceous schists. This gneiss is characterized by abundant slipping planes, which show movements parallel to the dip, and on which large plates of muscovite have often formed. As far as the observations showed, this gneiss is never crumpled and folded like the older gneisses.

The incipient gneissoid structure may be well observed in the gneissoid granite of Lost Horse Pass near Twin Lakes. (Pl. V.) Macroscopically it is light-gray granular rock of medium grain. The schistose structure is not prominent, but indicated by the distribution of the biotite and the lenticular shape of larger feldspar crystals. Under the microscope the large grains consist of quartz, orthoclase, and oligoclase, very similar to those in the normal granite. The biotite is largely recrystallized and converted to stringy aggregates of the same mineral partly changed into chlorite. Some muscovite is also present. There is an abundant development of new allotriomorphic fine-grained aggregates along winding and curving shearing planes. This aggregate consists of feldspar, mostly without striations, and quartz; there is also much micropegmatite, and in places a coarse poikilitic structure. Black iron ore in irregular, sometimes pressed, grains as well as apatite and zircon are accessories.

About halfway up Lost Horse Creek similar but more schistose rocks were observed. Their appearance under the microscope is similar to the specimens just described; in addition the larger grains are greatly crushed with much cataclastic structure. The normal gneiss from Mill Creek, 2 miles above the mouth of the canyon, is a plainly schistose rock with large orthoclase crystals pressed into partly lenticular shape. Biotite and a little muscovite lie in flat aggregates between streaks of pressed feldspar and quartz. There is much cataclastic action and formation of new allotriomorphic aggregates along wavy lines, which indicate schistosity. Large feldspar and quartz grains, when crossed by these lines, are greatly crushed. On the whole, however, the rock has great similarity to the normal Bitterroot granite.

Stronger crushing action near the mouth of the canyons simply results in emphasizing the characteristics already referred to. The sinuous lines along which most active recrystallizing movement is in progress multiply, and at last the whole rock is transformed into a fine-grained aggregate traversed by wavy streaks of finely distributed biotite and a little muscovite, and in this mass lie as pseudophenocrysts the irregular, more frequently lenticular remains of larger grains. These are often cut by parallel cracks which dip about 70° E., and which are sometimes opened and filled by secondary quartz, suggesting a stretching movement.

Still further pressure results in dark, fine-grained bands in the coarser gneiss, which under the microscope appear to be suffused by extremely finely divided biotite with allotriomorphic feldspar and quartz. This mass contains small, rounded residue of larger grains, and the whole imitates fairly successfully a fine-grained clastic rock.

The second gneiss area is that of St. Marys Peak in the northern center of the range and probably underlying the sedimentary Lolo series. This dark-gray gneiss is very rich in black mica and weathers to reddish outcrops. It is usually crumpled and includes portions or streaks of exceedingly micaceous schists. The dip is usually steep toward the west. The origin of the rock is uncertain.

On the east side of the Bitterroot Valley the granite, which is presumably but not certainly of the same age as the normal Bitterroot granite, contains frequent, well-defined inclusions of a crumpled gneiss, like that of St. Marys Peak. It is believed that this gneiss is composed of fragments of a more extensive formation which had been shattered by the granite. In places the granite itself is somewhat schistose. Dip and strike of these roughly gneissoid rocks, as well as those of the inclusions, are characteristically variable.

The next big gneiss area is found at the very head of the South Fork of the Bitterroot, adjoining the quartzitic series of that valley, which probably is identical with that of the Lolo Fork; the contact, though difficult to interpret, probably indicates that the gneiss underlies the sedimentary formation. This gneiss, which is cut by many granite dikes, consists of medium-grained aggregates of quartz and feldspar with some larger deformed crystals of the same kind. The abundant aggregates of muscovite and biotite run in irregular bands, curving around the lenticular patches of quartz and feldspar. It is often contorted and sometimes of coarse grain. It contains much microcline and little plagioclase. The schistosity is variable, but most frequently it dips 40° E.

A similar gneiss area, perhaps connecting southward with the one just described, is found on Salmon Mountain; it is extensively broken and intruded by granite dikes. Both these areas are pretty surely dynamo-metamorphosed granitic rocks of much greater age than the Bitterroot granite.

A much larger gneiss area is that of Elk City. It is elliptical in form and extends 40 miles north and south and 24 miles east and west. It borders almost everywhere against granite, the contact plainly showing the intrusive nature of the latter rock. The gneiss of this area is uniform in character; it is of light-gray color, roughly schistose, and consists of a granular mass of white feldspar and quartz, occasionally with larger orthoclase crystals; the aggregates of biotite indicate the schistosity. Microscopically the rock was found to consist of quartz, much microcline, grains of oligoclase or andesine, together with much black mica and sometimes a little hornblende. Crumpling and folding perpendicular to the schistosity are usually characteristic. Intruded narrow dikes of granite or pegmatite have often suffered the same crumpling. This rock also is clearly a dynamo-metamorphic form of a very old granite. The dip is generally flat, but characteristically variable.

An area of gneiss with equally variable strike and dip and also adjoined by intrusive granite begins at Pierce. Similar rocks occur in the lower part of the canyon of the North Fork of the Clearwater and at Oro Fino, along the main stream. Mica-schists, sometimes garnetiferous, and (at Oro Fino) limestone beds also occur with gneissoid rocks and the series is probably made up partly of highly altered sedimentary rocks, partly of sheared intrusives, both older than the Bitterroot granite. The intense and equally distributed metamorphism of this series is probably not due to the adjoining granite, but rather to regional causes.

TERTIARY LAVAS.

The distribution of younger igneous rocks in this region is peculiar. A line of rhyolitic eruptions of moderate volume follows the Bitterroot Valley. No basalts or andesites are known to exist here. Within the whole extent of the Bitterroot and the Clearwater Mountains no surface lavas of any kind have been found, but at the foot of these same mountains is spread out a veritable flood of Miocene basalts, covering thousands of square miles. Neither rhyolites nor andesites are known from this part of the country.

Rhyolite.—Beginning at the north the first small area of rhyolite is found a few miles west of Florence in the Bitterroot Valley, forming a low ridge near the first foothills. Forty miles farther south a much larger area is exposed in the bluffs on the east side of the river, forming a succession of tuffaceous, vitreous, and felsitic flows several hundred feet thick. This rock breaks through the granite and gneiss of the eastern foothills and many long dikes in the vicinity indicate its manner of eruption.

Dikes of granite-porphyry, many of which have a somewhat rhyolitic habit, occur along the South Fork above the junction. Between Blue Joint and Overwhich creeks, well up toward the head of the stream, another not very thick rhyolite flow is met with, a small area still remaining on the summit of Crown Peak or Castle Rock south of Nez Perce Pass. It filled the bottom of the valley to a depth of several hundred feet. In its upper part sedimentary beds and lignite seams are intercalated in the felsitic flows. Still higher up near the head of the river, on Lick Creek, a succession of flows of glassy, felsitic, and tuffaceous rhyolite is met with; this came down from some point on the divide between the South and East forks.

While the general habit of these rocks unquestionably is that of rhyolites, the general prevalence of phenocrysts of oligoclase suggests that they contain a a large amount of soda. No analyses have been made. The age is probably Miocene or Pliocene.

Basalt.—From the last heights along the western margin of the Clearwater Mountains the vast plateau of the Columbia River basalt, covering thousands of square miles, spreads out in a magnificent view. The old Clearwater Plateau, rising to 7,000 feet above the sea, heavily forested, and deeply trenched by canyons, breaks off with rapid slope, and at its foot extends another plateau, much better defined, 4,000 feet lower, but still 3,000 feet above the sea, its open fields in strong contrast to the dark-green Clearwater forests. The canyons are in reality much sharper incised than those of the older plateau, but seen from this point of view become indistinct. With a knowledge of the geological structure of the lava plateau, we can easily imagine how the molten rock in rapidly succeeding flows swept over and buried the foothills, perhaps at most permitting isolated high points to project above the lava, until at last these horizontal basaltic sheets had frequently attained a thickness of 2,000 feet and in a few places even of 4,000 feet. The enormous volume of congealed magma along the foot of the mountains dammed the old canyons of the rivers and forced the streams in their lower courses to seek new channels. New drainage systems were laid out on the basalt plateau; and corrading rivers cleared the partly filled canyons of the damming lava and excavated new trenches in the basaltic fields.

The basalt is a normal rock of its kind, usually an olivine-basalt, with a tendency to diabasic granular structure, glass remaining between the grains. Each of the many individual flows is ordinarily scoriaceous on top and more massive in the center. In a few places waterlaid clays and sands, with local lignites, are intercalated, doubtless accumulated by the rivers during lulls in the eruptions. Beds of tuff are absent.

The age of this basalt is generally conceded to be Miocene. The evidence, consisting mostly of fossil plants in interbedded sediments, has been gathered from various points and examined, chiefly by Professors Knowlton and Merriam.[a] No new evidence as to the age of this basalt has been obtained during this reconnaissance.

The original surface of the basalt was no doubt approximately level over large areas; still we are not justified in applying hydrostatic laws to the fluid rock, and

---

a Knowlton, F. H., Bull. U. S. Geol. Survey No. 108, 1893, pp. 103–104.

it is very likely that if a focus of eruption was located high up on a mountain side the flows might have an initial dip of several degrees down to the level at which the main valleys were flooded. No positive evidence as to the manner of eruption has been obtained. As a negative bit of evidence it may be stated that no basalt dikes of any kind have been found along the western margin of the Clearwater Mountains. The only focus thus far discovered along the eastern shore of the basalt plain is that described in a previous publication[a] and located at elevations of up to 8,000 feet on the eastern slope of the Eagle Creek Range, in Oregon. As has been surmised from the great fluidity of the lavas, the magma seems to have been emitted from a great number of short dike fissures in quiet and continuous streams.

Orographic dislocations have caused two folds in the level surface of the lava plateau. Most important among these is, first, the synclinal depression centering in Lewiston and bordered on the north by the monoclinal fold of the Clearwater escarpment; and second, the gentle anticline of the Craig Mountains, separating the Lewiston Plateau from Camas Prairie.

Slow earth movements on a grand scale are also believed to have affected the lava plateau. The principal movement seems to be a universal depression, perhaps induced by the extrusion of vast basalt masses from the interior. From a study of the courses and canyons of the antecedent rivers, like the Snake and the Salmon, it is surmised that certain marginal portions of the plateau may have been raised above the original level and that the extraordinarily deep canyons have resulted from an erosive trenching progressing simultaneously with this uplift.

QUATERNARY FORMATIONS.

The distinctly Quaternary deposits of this region are principally those connected with glaciation. Heavy moraines lie at the foot of the southern part of the Bitterroot Range, and glacial accumulations are found in all the valleys descending eastward and westward from their crests. From the Nez Perce to the Lolo Pass this range was covered by a sheet of névé and ice. Smaller moraines indicating a local glaciation are found on Salmon Mountain, The Crags, Buffalo Hump, Grave Peak, Rhodes Peak, and the Rocky Ridge, but over a larger part of the Clearwater Mountains below an elevation of 7,000 feet, as well as along Lolo Fork and the South Fork of the Bitterroot River there is no indication of glaciation. It is also believed that the larger part of the Coeur d'Alene Mountains was free from ice during the Glacial epoch. South of Mullan Pass and generally near the highest points there is some evidence of local glacial basins. In this more northerly region the limit of glaciation seems to lie a little lower than in the Clearwater Mountains.

[a] Lindgren, W., Twenty-second Ann. Rept. U. S. Geol. Survey, pt. 3, 1902, pp. 740–742.

Deep gravels accumulated in the valley of the Bitterroot River and its branch streams during the Glacial epoch and, since its close, have been gradually excavated and terraced. The bottom lands along the rivers and the dried-lake basins in the glaciated district are the most recent Quaternary deposits.

GEOLOGICAL HISTORY.

An exact statement of the geological history of this region is difficult to give on account of the few exactly determinable datum planes. There are really only two determinations of time on which we may rely. The first is the date of the Columbia River lava as Miocene; the second is the, Glacial epoch as early Quaternary. Back of the Columbia River lava the ages given for the sedimentations, intrusions, and dislocations are only tentative.

The oldest rocks known are probably the gneisses of the Clearwater Mountains; more recent than these are the sediments of supposed pre-Cambrian, Carboniferous, and Triassic age. We may conceive a post-Triassic uplift of great importance, followed or accompanied by vast intrusions of granite, the granite of the Bitterroot, Clearwater, Salmon, and Coeur d'Alene mountains.

We may further, with great probability, assume a long-continued erosion, which planed down this uplift to the moderate relief of the Clearwater, Salmon, and Coeur d'Alene plateaus, and which exposed the intrusive granite by the removal of great masses of covering sediments.

Going one step further, it is necessary to assume a second great and evenly-distributed uplift which raised this eroded surface several thousand feet above the sea, an uplift of post-Triassic and pre-Miocene age. This was probably accompanied by breaks along what is now the western margin of the plateau, and was necessarily followed by the establishment of the systems of the Bitterroot, Clearwater, and the Salmon rivers. These streams trenched the plateau broadly and deeply, with the result that their canyons during the Miocene epoch were cut to a depth equal to that of the present day. The relations of the Columbia River lava to the old topography prove this on the western side, and similar evidence is adduced for the South Fork of the Bitterroot, by the fact that the rhyolites there filled a valley coinciding in depth and configuration with that of to-day.

Going back a little further from the Miocene datum plane, perhaps to a time when the Bitterroot River first flowed over the uplifted plateau, which probably reached far into Montana, a dislocation of great importance occurred along what is now the Bitterroot Range. This dislocation, the beginning of which perhaps scarcely antedated the close of the Cretaceous period, extended for 60 miles north and south, only bending westward close to its southern end. This fault plane was inclined eastward at angles up to 26° from the horizontal, and the rocks along it bear evidence both of molecular and molar movement;

the former expressed in schistosity, the latter in striated slipping planes. The direction of the movement was that of a normal fault, and seemed to have been of a stretching and shearing character. The foot wall seems to have moved up as expressed by the raising of the Bitterroot Range above the general level of the Clearwater Plateau. The hanging wall seems to have moved down as expressed by the apparently structural trough of the Bitterroot Valley. The minimum amount of the dislocation along the plane of fault is 2 or 3 miles; the minimum horizontal component would be but little less, while the corresponding vertical component is about 5,000 feet. The evidence finally shows that movement along this dislocation has proceeded for a very long time, and probably still continues along certain parts of the fault. The flat dip of the fault plane has naturally aided the preservation of the record.

During the latter part of the Miocene epoch, important events happened. The western foot of the Clearwater Mountains was flooded by basalt, poured out through fissures in enormous volume. The foothills were covered by rapidly succeeding outbursts of very fluid lava, piled up in a great number of thin flows, which had a maximum thickness of 4,000 feet, and finally reached a level of about 3,000 feet above the sea, as the land stands to-day. The lower river courses were entirely changed; while the canyons of the Clearwater Mountains, though dammed by basalt, still preserved their form and direction.

The last part of Tertiary time, and perhaps the beginning of the Pleistocene, was occupied by the rivers on the western slope in undoing the constructive work of the earlier eruptions. Steep and sharply incised canyons were eroded along new stream courses along the lower Clearwater and the Salmon rivers, while the trunk stream of the Snake River cut out an entirely new course from Wieser (at the lower end of the Great Snake River Valley) to its mouth. The canyons were cleared of accumulations, and at the beginning of the Glacial epoch the rivers of the western slope had trenched their canyons very nearly to the depth they have to-day. The filling of the canyons by lava had probably, in most cases, produced accumulation of gravel terraces along the upper parts of the watersheds, and of such character are probably the auriferous terrace gravels of Elk City, on the South Fork of the Clearwater, the highest level of which reaches 4,500 feet in elevation. In the lower stream courses the pre-Miocene drainage channels now lie below the bottom of the present streams; from this it seems probable that a large part of the field of the Columbia River lava has undergone a subsidence since the time of the eruptions; a subsidence probably caused by the eruption of large masses of magma from the interior of the earth. Minor folds have occurred in the once horizontal lava beds, as near Lewiston. Finally, the courses of the Snake River and the lower part of Salmon River seem to indicate that an uplift has occurred along certain marginal parts of the lava plateau; the deepening of the canyon keeping step with the rate of the uplift. Only on such a supposition can we account for the abnormally deep canyons, which sometimes are cut 5,000 feet below the plateau surface.

On the eastern slope the happenings of late Tertiary time are less easily traceable, but it is probable that the damming by basalt of the Clark Fork of the Columbia and the subsidence of Bitterroot Valley along the great fault made a lake basin of this valley for a limited time.

One of the last events in the geological history is the glaciation which during the later part of the Quaternary period covered the whole Bitterroot Range with ice and snow and produced the extremely rugged forms which now characterize its surface region. Glaciers extended to the mouth of the eastern canyons of the range and also far down into the Clearwater drainage. Great moraines were accumulated at the southern end of the Bitterroot Mountains. The Clearwater Mountains were not uniformly glaciated, but contained numerous small glacial centers, such as The Crags, Grave Peak, Rocky Ridge, Salmon Mountain, and the Buffalo Hump region. From causes not clearly recognized as yet, the Lolo Fork, the main valley, and the South Fork of the Bitterroot River were during the Glacial epoch filled with gravels to a height of 400 or 500 feet above the present level of the stream. Similarly, we find evidence of a damming of the Snake and the Salmon canyons, with accompanying accumulation of sands and gravels to a depth of about 300 feet. Less evidence of this kind is available from the Clearwater system, though this also must have been similarly filled.

Finally these accumulations were removed, and during this clearing out the terraces and bottom lands which now line the Bitterroot Valley were formed; gravel bars were left in places along the lower Clearwater, the Salmon, and the Snake; morainal material was worked over and terminal moraines trenched; basins of glacial lakes were filled and converted into meadows.

Beyond this there has been only a slight deepening of the canyons of the great rivers. In glacial regions the marks of the ice seem as if dating from yesterday. Except in the region of the Columbia River lava, the rivers have deepened their channels but little since early Tertiary time. The influences causing this retardation are the Miocene lava dams, the post-Miocene subsidence, and the obstructions to the rivers during the Glacial epoch.

BITTERROOT VALLEY.

TOPOGRAPHY.

Bitterroot Valley extends, properly speaking, from the point where the Lolo Fork enters the Bitterroot River to a point 5 miles south of Darby, where the river splits into two main branches. It is a well-marked depression about 65 miles long and up to 9 miles wide in its broadest portion, with a nearly north-south direction. In the upper 15 miles its width is only from 2 to 4 miles. Westward the Bitterroot Range rises with its remarkably even slope and deeply incised canyons, while eastward, with gentler slope and far less pronounced outlines, are the foothills which gradually lead up to the main divide of the Rocky Mountains toward Phillipsburg. The Bitterroot River pursues its course in a northerly direction with an average grade of 20 feet per mile, receiving many tributaries from the western mountains and a few from the much more arid eastern foothills. Those from the west, which generally have an east-west direction within the range, are usually sharply deflected northeasterly, where they debouch into the valley.

CULTURE AND VEGETATION.

Owing to a moderate elevation of from 3,900 to 3,150 feet, good soils, and excellent water supply, the Bitterroot Valley forms one of the most prosperous agricultural areas of Montana. There are many little towns, the largest among which is Hamilton, picturesquely situated in the middle of the valley, with beautiful view across to the serrated and snowy ridges of the Bitterroot. The hardier fruits thrive, as well as cereals and forage plants, but irrigation is everywhere necessary. Along the river is a moderate growth of deciduous trees and a few pines; the gravel terraces and the eastern foothills are partly open; while the lower hills of the Bitterroot Range are covered by thick timber. The forests and soils of this region have been described in much detail by Mr. J. B. Leiberg.[a]

GEOLOGY.

QUATERNARY.

A large amount of Quaternary deposits lie in the deep trough of the Bitterroot Valley, and there is evidence that it was once filled to a much higher level than at present. The greater part of this material is doubtless derived from the moraines and glacial streams which came down from the valleys of the western range, but much of it has also been brought down from the canyons of the upper river.

At the forks of the river, 25 miles south of Hamilton, the alluvial bottom lands are about 1 mile wide and continue with the same width downstream toward Grantsdale. From this point northward to the end of the valley the bottom lands are ordinarily several miles wide. These latest alluvial deposits are throughout the valley flanked by gravelly terraces, often several miles wide. Near the forks of the river and upstream toward Darby there are at least two prominent terraces on the west side at elevations of 100 and 200 feet above the river. In some places these are sharply defined and at others merge into the slopes of the valley. The highest is at an elevation of about 4,100 feet. Near the river

---

[a] Nineteenth Ann. Rept. U. S. Geol. Survey, pt. 5, 1899, pp. 253-282.

are lower gravel terraces, 8 and 30 feet above the water level. The upper terrace leans against gradually rising hills of heavy moraines. On the east side there are few terraces south of Grantsdale, for the river is closely crowded against the rocky foothills of that side. From Grantsdale north the terraces become wider. East of Roaring Lion Creek there is a 30-foot terrace on the west side of the river, with a gradual slope covered with heavy gravel wash, which extends to the first moraine at the mouth of the canyon, which is 400 feet above the river and at an elevation of 4,000 feet. Similar relations obtain at the mouth of Mill Creek, except that the gravel is finer and more sandy.

At various places between Curlew mine and the mouth of Bear Creek the gravel terrace contains much quartzite in large cobbles, the origin of which seems somewhat doubtful, as there is not much quartzite in place close by. At the Curlew mine the gradual gravel slope reaches up from the river to the foothills, 200 feet above. On a spur 500 feet above the mine and 700 feet above the river lies a patch of well-washed gravel, the elevation of which (4,033 feet) must indicate the height to which the valley was once filled. Immediately north of Big Creek the gravel again reaches up to an elevation of 4,000 feet.

Near Florence, in the northern part of the valley, the gravel terrace skirts the river, the banks being 20 feet high in places. From the Curlew mine north the gravel contains much coarse, micaceous gneiss and the sands are full of mica. The evidence, as far as it goes, indicates that during the Glacial epoch the Bitterroot Valley was filled with gravels and sands to an elevation of about 4,000 feet along the margins, and that the grade of the river was then less than at present.

No direct evidence of the former existence of a lake within the valley has been found. The east side, however, has not been carefully examined, and from the general configuration it would seem possible that this depression, like many other intermontane valleys of this region, was occupied by a lake in late Tertiary time.[a]

It seems highly improbable that the Bitterroot Valley, at least north of Grantsdale, has been formed by erosion alone. It bears too many characteristics of structural valleys, and it seems not out of the way to tentatively associate its origin with that of the great range to the west of it. This subject will be referred to again in the description of the mountains.

EASTERN FOOTHILLS.

TOPOGRAPHY.

As far south as Skalkaho Creek the foothills on the eastern side of Bitterroot Valley are low, ascending gradually in many low spurs from gravel terraces. "The crest line lies about 30 miles east of the valley, but as a continuous ridge

---

[a] Mr. Bailey Willis states (orally) that he found soft, clayey beds, probably of lacustrine origin, on the north side of Skalkaho Creek, in the first foothills, a few miles from Hamilton. These beds dip westward at an angle of 80°.

is scarcely so definite and well marked as the summit of the western watershed. The ultimate divide is the main range of the Rocky Mountains, but the intermediary ranges, radiating in all directions, form so many separate divides that the central backbone is lost sight of. The canyons that enter the valley from the east are not so numerous nor so regularly arranged as are the western affluents. The divides between them rise from the valley with easy gradients, presenting fronts many miles in width that are seamed and scarred with shallow and irregular gulches, sometimes cutting back several miles in the fronting portion of the spur."[a] South of Skalkaho Creek the spurs advance close to the river, presenting a steep front in places over 1,000 feet high. The irregular maze of closely crowded ridges gradually attains elevations of over 7,000 feet in the complex of hills a few miles east of Darby, which still is many miles distant from the main divide.

GEOLOGY.

SEDIMENTARY SERIES.

While the prevailing rock east of the river appears to be a granite or gneiss, there are also considerable areas of a sedimentary series, chiefly quartzite or calcareous quartzite. The age of this series is in doubt, as no fossils have been found, but there is considerable similarity to certain parts of the Lolo Fork series and the rocks may well be of the same age. As on Lolo Fork, the contacts with the granitic rocks indicate the intrusive character and later age of the latter.

In the foothills east of Victor the first outcrops of older rocks consist of compact quartzites dipping 27° to 48° W. Larger areas of quartzite appear in Skalkaho and Weeping Child creeks, a few miles distant from the valley. On Skalkaho Creek the beds dip 45° to 55° N. or NE. On Weeping Child Creek about 5 miles above its mouth is a narrow belt of quartzite, with a dip of 85° NW. This continues across to Skalkaho on the north and can be followed southwesterly across the spurs to a steep bluff on the Bitterroot River, a short distance north of Como post-office, where the dip is 38° SE. Small croppings of quartzite are even found in places below the gravel terrace on the west side of the river. Beyond the first belt of quartzite on Weeping Child Creek granite or gneiss is exposed for a distance of 2 or 3 miles; the quartzites then come in again with varying though generally flat dip, and near the Hot Springs are interstratified with micaceous schists. The series is evidently both folded and faulted.

CONTACTS.

On both creeks the contacts with the granitic rocks are well exposed and show that the latter are of unquestionable intrusive character. At the main contact on Skalkaho Creek the quartzite is magnificently intruded by branching dikes of

---

a Leiberg, J. B., Nineteenth Ann. Rept. U. S. Geol. Survey, pt. 5, 1899, p. 256.

normal granitic rock; near the granite the quartzite contains bands of biotite-schists and is largely converted into banded hornfels with malacolite and biotite. Inclusions of these rocks are also contained in the granite. In places the granite dikes are pegmatites with much quartz. At the first contact in Weeping Child Creek the same phenomena are seen. Many of the granitic dikes here follow the bedding planes of the quartzite, while in the section described above they mainly break across them. The same contact metamorphic rocks are noted.

GRANITIC ROCKS.

The larger part of the foothills east of the river is occupied by granites and gneiss, the relations of which seem rather complicated. Much of the granite is a normal, massive, light-yellowish gray rock similar to that of the central Bitterroot Range and, like that, containing both biotite and muscovite. The grain is apt to vary and some specimens contain large orthoclase crystals. Such rocks are found on the lower Skalkaho and southeast of Darby. Larger and smaller fragments of gneissoid rock, rich in black mica and with contorted planes of schistosity, frequently occur as inclusions in this granite. Over large areas the granite itself has acquired a roughly schistose structure. This is seen on lower Weeping Child Creek, where the schistosity has an east-west direction, and in the hills east of Darby, where it strikes from N. 15° W. to N. 50° E., the dip being either westerly or easterly from 45° to 60°.

The most probable interpretation of these facts is that an old pre-Cambrian series of quartzite and of still older true gneisses was intruded by granite similar in age and type to the Bitterroot granite. While large quartzite areas are left, the contorted inclusions are the only remnants of the gneisses. The granite has subsequently by irregularly acting forces acquired local rough schistosity in various directions. The included gneiss fragments can not reasonably be interpreted as contact metamorphic parts of the quartzite series. The roughly gneissoid granites are very different from the strongly compressed gneisses of the Bitterroot Range and probably did not result from the same kind of stresses which were active at that place.

DIKES.

The granitic rocks contain abundant dikes of granite-porphyry, dark diorite-porphyry, and rhyolite.

RHYOLITE.

For a few miles south of Como post-office the east side of the river up to a height of 1,000 feet is lined by precipitous bluffs which are formed by a series of rhyolite flows dipping about 45° to 60° SE. A thickness of several hundred feet is exposed and the succession from the base up is as follows, the flows resting on gneissoid rock:

Section near Como post-office.

(1) Columnar rhyolitic tuff, weathering soft.
(2) White rhyolitic tuff, weathering soft.
(3) Red felsophyre, weathering brown.
(4) Soft white rhyolite.
(5) Massive white rhyolite.
(6) White breccia, with dark bands.
(7) Red felsophyre, with green and black bands of vitrophyre. Rhyolite is also found in a few places on the west side of the river covered by Quaternary terrace gravel.

LOLO FORK AND NORTH END OF BITTERROOT RANGE.

TOPOGRAPHY.

At Lolo post-office, 35 miles north of Hamilton and 10 miles south of Missoula, the foothills of the Bitterroot Valley advance close to the stream, and this point may be considered as the end of the valley, properly speaking. Near the same place is the northern termination of the Bitterroot Range. From Lolo Peak, about 9,000 feet in elevation, the slopes descend abruptly toward Lolo Fork, an important tributary of the Bitterroot River; north of Lolo Fork the complicated system of timbered ridges which mark the beginning of the Coeur d'Alene Mountains rarely attain 7,000 feet above the sea.

Along Lolo Fork for a few miles above its mouth the bottom lands have a maximum width of one-half mile, and are bordered on both sides by gradually rising hills. Soon, however, a canyon begins with flat though narrow bottom, and this is practically continuous to the head, the slopes rising steeply on both sides for a few hundred feet, continuing then along more gently inclined ridges. About 20 miles from the mouth an important tributary is received from the north, and near this place the valley widens and contains some meadow land at an elevation of 3,700 feet. A well-marked terrace is noted 500 feet above the valley, and from this the long ridges gradually ascend to elevations of 6,000 feet or more. An excellent view is afforded from them southeasterly toward the snow fields and serrated peaks of the Bitterroots, and northwesterly over the monotonous, heavily timbered ridges of the Coeur d'Alene Mountains, scarred by the tracks of extensive forest fires.

Above the North Fork the canyon continues for 10 miles to Hot Springs where the stream forks again, one branch extending 10 miles westward and heading near Rhodes Peak, while the other comes down with northerly direction from Lolo Pass, 8 miles distant.

Lolo Fork thus runs from west to east just north of the Bitterroot Range and may be considered to mark the boundary between this range and the southern part of what, in this report, has been called the Coeur d'Alene Mountains. Its length is about 40 miles, and its southern fork cuts in back of the main divide of the Bitterroot Range. A wagon road leads up to the Hot Springs (elevation 4,100 feet), a favorite summer resort with baths, and at this place the Lolo trail begins. In 1899 this part of the trail was in very bad condition, due to fallen timber between which the path runs in most eccentric curves. A gradually ascending ridge leads up to the divide which has an elevation of only about 5,200 feet. Lolo Pass is a broad, flat depression between the westerly spurs of the Bitterroots and the projecting ridges of Rhodes Peak in the Clearwater Mountains. Deep soil and extremely thick forest with underbrush cover the hills, between which extend a series of swampy meadows. A few miles westward the trail crosses the North Fork of the Lochsa Fork and views are obtained over the Clearwater Mountains, heavily forested, with deep though not precipitous canyons and with slightly undulating skyline. In many places white exposures of granite gleam in the openings of the forests.

GEOLOGY.

SEDIMENTARY SERIES.

For a distance of 20 miles in a straight line westward from its mouth Lolo Fork cuts across a sedimentary series with a general northwesterly strike and prevailing northeasterly dip. Near Hot Springs this series is cut off by intrusive granite, the contact line thence curving southward toward Lolo Pass. This same series occupies the extreme northern slopes of the Bitterroot Range and probably also Lolo Peak. Along the eastern front of the range it reaches down to the mouth of Carlton Canyon, 7 miles south of Lolo Fork, where it again is cut off by intrusive granite; thence the contact probably extends westward toward Lolo Pass. The micaceous, contorted gneiss of St. Marys possibly connects with the sedimentary area.

These sedimentary beds, which may be provisionally designated as the Lolo series, derive additional interest from their probable identity and continuity with the series occupying the entire width of the Coeur d'Alene Mountains, 90 miles northward, and extending from the lake of the same name on the west to Clark Fork of the Columbia on the east. In neither case has the age been established, no trace of fossils having thus far been found. That the beds are very old is evident; they may even be pre-Cambrian.

The study of the series began at the mouth of Lolo Fork close to the Lolo post-office. Cherty limestone and quartzite crop here with northwesterly strike, adjoined 1 mile farther up the fork by a narrow belt of purple shale or slate. Three miles above begins the mouth a broad belt of quartzite-schist with northwesterly strike and a dip of 83° NE. Going up Mormon Gulch, which empties into the fork near its mouth and heads near Lolo Peak, shales and cherty limestone are noted; these beds have a southwesterly dip of 55°; above them on the same gulch begins the same micaceous quartzite-schist which was observed on the main fork and which contains a few aplitic dikes.

In the main fork the quartzitic schists continue for several miles. The schist, which is clearly of sedimentary origin, splits easily in slabs coated with muscovite. Just below Westerman's ranch, 5 miles from the mouth of the river, a belt of thick-bedded quartzites crosses the canyon. The next part of the series continues for 2 miles and consists of calcareous quartzite with some shaly limestone, with same northwesterly strike and northeasterly dip. Half a mile above Woodmans Creek this is adjoined by heavy bedded quartzite, continuous for 1½ miles. The canyon now becomes narrower, and schists of a greenish color, probably largely altered eruptives, predominate. One mile below Spark's ranch a dike of quartz-diorite crosses the canyon; above this a white quartzitic schist is extensively developed, striking N. 44° to 74° W., and dipping 35° to 45° NE. A long ridge north of Spark's ranch was ascended to a point 2,000 feet above the valley and the series was here found to strike much more northerly, ranging from due N. to N. 15° W., and dipping about 20° W., implying some kind of a break between this part and the rocks exposed in the canyon. The beds are banded, calcareous rocks, greenstone-schists, and a large amount of purple slates, very similar to those near the mouth of Lolo Fork.

The white quartzitic schists continue with the same dip for several miles above Spark's, when they run over into cherty banded and streaked calcareous quartzites which do not appear very much altered. In spite of careful search no definite fossils were found. One dike of diorite-porphyry was observed on the north side of the river. The remaining distance of about 6 miles up to the granite contact near Hot Springs is occupied by banded and calcareous quartzites, interstratified with dark slates. While this series contains lime, it is not so distinctly calcareous as that exposed at the mouth of Woodmans Creek. The strike swings more to the west, being from N. 60° to 80° W. The dip is prevailingly from 15° to 45° NNE. In a straight line the distance from the mouth of the fork up to the contact is 20 miles, but except in the lower part the river does not cut squarely across the strike. No estimate of thickness of this series can be attempted on the basis of the foregoing observations, but it is apparent that it must be considerable, even allowing for folding, faulting, and repetition. There is a noticeable absence of pyritic impregnation, quartz veins, and dikes. While the series contains some sheared eruptives, there is certainly no great amount of this kind of rock.

LOLO CONTACT.

Just below Hot Springs, near the mouth of Granite Creek, is a sharp contact, though the actual contact plane is not well exposed and the drainage basin above this point is occupied by granite. The contact crosses Granite Creek near its mouth, and thence for many miles continues with a general direction of N. 80° W. To the south the contact changes its direction and appears to follow pretty close to Lolo Fork. Near Lolo Pass irregular areas of metamorphosed sediments were observed. The thick vegetation and deep soil give few opportunities for study of the rocks.

For several hundred feet from the contact the quartzitic slates are very crystalline, and minerals and structure typical of contact phenomena have developed in them. The exposures on the ridge east of the Hot Springs are not good, but at the contact the slates are very micaceous. The calcareous rocks have been altered to malacolite-hornfels, while the pure quartzite has obtained a dense flinty structure. A granite-porphyry with bipyramidal quartz crystals often appears on the immediate contact. The excellent outcrops, 1,200 feet below Hot Springs in the creek, show greenish crystalline schists, striking north-south and dipping 80° W. These schists contain many small dikes of muscovite-granite, aplite, and pegmatite, some containing tourmaline. A specimen 20 feet from the actual contact which is covered shows a biotite-andalusite hornfels with abundant tourmaline. The alteration decreases rapidly, though a specimen collected 1,000 feet below the contact, on Lolo Fork, appearing like a dark-gray, flinty quartzite, still shows the structure of hornfels and contains tourmaline. Half a mile below the contact banded slates appear with normal strike and dip, and very little altered, the chief evidence being little spots and knots on the cleavage faces. A few dikes of granite-porphyry were noted at this place.

The intrusive character of the granite into the Lolo sedimentary series is the main point established.

CARLTON CREEK CONTACT.

As might be expected, the quartzitic schists and quartzites of the lower Lolo Fork continue with northwesterly strike across the ridge between that river and the main Bitterroot Range, and are found along the foothills for 7 miles south of the junction. Quartzite is exposed to a point three-fourths of a mile north of the mouth of Carlton Creek Canyon, where knotty slates are interbedded and show strikes of N. 45° W., dip 45° NE. The slates are micaceous and contain in places abundant narrow dikes of pegmatitic and aplitic granite 5 to 6 inches wide, chiefly intruded parallel to the coinciding planes of schistosity and bedding. Going southward, more quartzite follows, and at Carlton Creek the contact of the sedimentary series with the granitic is reached. One of the interesting points of this contact is that the intrusive rock here consists of the normal gneiss of the Bitterroot front slopes, itself a dynamo-metamorphic development of granite. At this place the gneiss is imperfectly schistose, but the dip is from 15° to 18° E., as at most points farther south. About 2 miles above the mouth of the canyon this gneiss seems to gradually change into the normal granite abundantly exposed on the slopes facing north. The exact place of contact between gneiss or granite and quartzite is somewhat in doubt, so gradually increasing is the injection of the latter by pegmatite and granitic rocks. That this contact, too, is intrusive admits of no doubt, and it is accompanied by a more intense contact metamorphism than at Hot Springs on Lolo Fork, probably, because the series is here fractured across the bedding planes. In some places the interstratified clay slates are converted to gneissoid mica-schists. The zone of metamorphism extends at least three-fourths of a mile northward from the contact.

GNEISS OF ST. MARYS PEAK.

Eighteen miles south of the northern end of the range, St. Marys Peak rises to an elevation of 9,300 feet or 6,000 feet above the level of the valley. As usual in these mountains the culminating points are situated a few miles east of the main divide. The long ascent from the valley leads over the usual sheared granite of the front of the range, with its constant easterly dip. But at about an elevation of 7,000 this changes without clearly exposed contact to a very contorted and entirely different gneiss, very rich in black mica and weathering with reddish color. Streaks and patches with fairly well-marked outlines of fine-grained, also contorted, mica-schist are embedded in this. These rocks have, on St. Marys Peak, a northerly strike and a westerly dip of about 70°. The area clearly extends southward across Big Creek. As evidenced by the detritus brought down to the valley by the creeks, similar rocks also occupy considerable areas to the north of the peak. A short distance westward the normal granite begins, and is well exposed in the glaciated outcrops. The age and origin of the St. Marys gneiss has not been determined. A possible explanation is to consider it simply as an extreme facies of contact metamorphic action on the Lolo series, and this view receives some support from the fact that small amounts of somewhat similar rocks were observed at the Carlton Creek contact. But the area is too uniform and large and the absence of quartzites is too conspicuous to allow this explanation to stand. Considering the relations in the southeastern and southwestern parts of the Clearwater Mountains, it is more probable that it is an independent and older gneiss, very likely of Archean age, which underlies the Lolo series and which, like those beds, was invaded and ruptured by intrusive granite.

GRANITE; QUARTZ-MONZONITE.

A granite of the normal appearance of the Bitterroot variety occupies the drainage basin of Lolo Fork, above Hot Springs, and of Granite Creek. The rock which near the Springs forms light-gray, large, partly rounded, outcrops of typical form, is coarse grained and contains both biotite and muscovite. An analysis would probably bring out its close relationship to quartz-monzonite. It contains much grayish orthoclase, besides perthite and a little microcline as well as an acid oligoclase. The character seems constant as far as observations were extended. A few miles west of Lolo Pass a few outcrops of pyroxenite were noted; beyond this the normal muscovite-granite begins again, and from the appearance of the outcrops, seems to occupy the whole of the last end of Lolo ridge in the Clearwater Mountains and the larger part of the country visible to the south of the Lochsa Fork, including the prominent mass of Grave Creek.

QUATERNARY.

The only deposits belonging to the Quaternary period, noted in the Lolo drainage, are the narrow meadow lands appearing at intervals in the usually flat bottom of the Lolo Fork and the high terrace near Sparks's ranch. While in the lower part of the stream no gravel terraces were observed, an extensive bench of coarse gravel, deeply cut by the river and its gulches, forms a prominent feature near the junction of the first important branch from the north, 14 miles above the mouth of the river. Steep declivities lead up from the bottom lands to this terrace, which probably occupies an area of several square miles on both sides of the river. The well-washed gravel begins at an elevation of 400 feet above the stream and reaches up to 550 feet (elevation 4,300 feet); above it rise long, gently sloping ridges of slate and other sedimentary rocks.

This occurrence is interesting as it shows that during some part of the Quaternary period Lolo Fork, like the main Bitterroot River, was filled by sands and gravels to an elevation several hundred feet above the present river bottom.

No evidence of glaciation was noted along Lolo Fork, though it is probable that the ice streams of the Bitterroot Range extended westward to a point near Lolo Pass and that the meadow lands at that place are of glacial origin.

BITTERROOT RANGE.

TOPOGRAPHY.

As defined in this report, the Bitterroot Range extends from the head of West Fork of Bitterroot River and the Nez Perce Pass on the south a distance of 65 miles almost due north to Lolo Fork. Its imposing front, with deep canyons and snow-flecked serrated ridges, overlooks the Bitterroot Valley on the east, while westward its slopes soon merge into the dissected high plateau of the Clearwater Mountains. Its height is considerable, though many ranges of the west easily exceed it in this respect. Near the southern end El Capitan and Trapper Peak nearly reach or exceed 10,000 feet, but the average elevations of its highest summits is more nearly 9,000 feet. These elevations are very constant throughout the range, which thus rises somewhat less than 6,000 feet above Bitterroot Valley. North of Lolo Peak and south of Boulder Peak the elevations decrease conspicuously, few points exceeding 8,000 feet south of the Boulder Peak or 7,000 feet north of Lolo Peak. The range is characterized by some very singular and constant topographic features, of which the most striking, perhaps, is the regular and even slope toward the valley. From Lolo Peak to Nez Perce

FIG. 2.—Bitterroot Range, looking northwest from Hamilton; showing gentle eastward dip of gneiss zone and deeply incised canyons of Blodgett and Mill creeks, with precipitous sides.

Pass this slope is nearly constant at angles ranging from 18° to 26°. Its width varies from 2 to 4 miles. Pls. IV, VI, B, VII, B, IX, A, and X and figs. 2 and 3 show the appearance of this slope from various points in the valley. From the summit of this slope, east-west ridges extend toward Clearwater to the divide which forms the boundary line between Montana and Idaho. These ridges are usually narrow, extremely jagged, and rough, but their highest peaks do not vary much in elevation. In some places, as near Ward Peak (Pl. III), the highest elevations of the range are nearer to the valley than to the main divide. Ordinarily, however, the highest summits of the range are halfway between the valley and the divide, the peaks along the divide never reaching 9,000 feet and ordinarily ranging in elevation from 7,500 to 8,500 feet. The saddles on the divide are from 7,000 to 7,500 feet in elevation, except in the case of the gap at the head of Lost

Horse Creek, which sinks to 6,600. Both Lolo Pass and the Nez Perce Pass are still lower, the former only attaining 5,300 feet. Thus the summit of the range lies several miles to the east of the Clearwater divide. (See profiles, Pl. IV.)

Another remarkable feature is in the direction and character of the drainage. Seen from any convenient place in the valley, for instance, from Grantsdale, the bold, even slope sweeps along the face of the range as a single tectonic plane; the canyons, like those of Blodgett and Mill creeks, appear as deep gashes, scarcely interrupting the continuity of the front. They are narrow, especially near the mouth, the walls in many cases being abrupt precipices. About twenty canyons, closely spaced, score the east slope of the Bitterroot Mountains. In most cases their streams are remarkably straight, flowing from west to east, without important

FIG. 3.—South end of Bitterroot Range from point near Darby; showing gradual eastern slope of gneiss zone south of Tincup Creek. Moraines are shown in foreground.

branches, and separated by almost equally straight ridges, with narrow, jagged summits. Especially fine instances of this arrangement are furnished by the four or five creeks west of Hamilton. (Pl. III.) Broadly U-shaped in their upper courses (Pl. VII, A), due to energetic glacial action, they generally narrow to V-shaped incisions near the mouth. Near the head we note a general tendency to turn southward. By far most of their tributary gulches are received from the south, while the slopes to the north of the canyons generally present unbroken fronts. Lost Horse Creek is rather an exception to the rule, as it splits, not far from the mouth, into three branches. Along the main divide and sometimes also on the secondary east-west ridges small glacial lakes are found in abundance.

Jointing of the rock, erosion, and glacial action contribute to an excessive roughness of topographic detail, making the range one of the most inaccessible. Continuous travel along the divides is impossible, on horseback or on foot. In the creek bottoms, treacherous, swampy areas, a slender but impenetrable growth of lodgepole pine, tangled underbrush, and fallen logs make travel difficult, even along the few trails which cross the range or ascend the gulches. Between Lolo and Nez Perce passes the only safe horse trail across the range is that leading up Lost Horse Creek, and even this is far from being an easy one.

Ordinarily mountain ranges adjoining thickly populated valleys abound in trails or even roads, and are well known and frequented by the inhabitants of the plains. But the Bitterroot Range is different; its lofty summits are without the attractions of flowery meadows and easy paths; only the hunter and the explorer penetrate its wilderness. The lower slopes are covered by thick forests which gradually become more sparse as elevations increase and the white granite begins to gleam between the dark pines; finally vegetation almost ceases and the summit region spreads out with a maze of peaks and broken ridges of dazzling bright granite, flecked by still more brilliant snow fields. (Pls. VII, B, and VIII, A.)

The main divide, distant from 8 to 15 miles from the first foothills, is, as stated above, a succession of sharp, craggy peaks alternating with deep saddles at the head of the large canyons. Especially in the southern portion of the range it is irregular, swinging from east to west with curves the radius of which is from 1 to 2 miles. On a smaller scale the same thing is repeated on the secondary east-west divides of the eastern slope, the secondary canyons cutting deep recesses in them, narrow, precipitous of grade in their lower part, and opening into wide glacial cirques near their head.

The western slope of the range is even more rugged than the eastern. "The immediate slopes from the crest are here very abrupt, are cut up by immense gorges, and abound in precipices and extensive rock slides to such a degree that they are entirely impassable."[a] Westward these ridges soon descend to the level of the dissected high plateau of the Clearwater Mountains, and the glacial cirques contract to the winding, narrow, and inaccessible canyons of that system.

GEOLOGY.

SEDIMENTARY SERIES.

Except near its extreme northern end the Bitterroot Range is almost exclusively built up of granite and gneiss. The Lolo sedimentary series and the St. Marys gneiss have already been described in the preceding section. South of Carlton Creek no sedimentary rocks are known from the interior mass and only a few smaller areas have been observed along the first foothills.

[a] Leiberg, J. B., Twentieth Ann. Rept. U. S. Geol. Survey, pt. 5, 1900, p. 319.

The largest sedimentary area is that near Curlew mine; it occupies less than 2 square miles and really forms foothills projecting beyond the normal line of the base of the range. It begins just south of Big Creek and continues southward for 2 miles, having an average width of half a mile. The rocks, which are not very well exposed, are chiefly quartzite, or calcareous quartzite, in places converted into a fine-grained, dark-brown, contact-metamorphosed hornfels. Throughout the rocks are exceedingly crushed and broken, and rarely show their strike and dip. At one place, on the ditch toward Big Creek, was noted a dark lime shale with a southwesterly dip of 60°. On the west the contact with the gneiss runs nearly straight north and south and is a fault plane. A dislocation occurred along this line in 1898 and is still visible on the surface of a gently sloping ridge. It extends for 1,500 feet and is partly an open fissure, the downthrow being on the eastern side and amounting to 1 or 2 feet.

No well-defined sedimentary rocks have been found along the foothills for a long distance south of this area. Near the mouth of Sawtooth Gulch fine-grained rocks occur interbedded with the ordinary gneiss, but they seem only to be varying dynamo-metamorphic developments of granites of slightly differing texture. Between Chaffin and Tincup creeks, near the south end of the range, extends a short ridge parallel to the trend of the mountains and of somewhat complicated structure. Much of it is made up of a fine-grained, almost flinty rock, largely consisting of quartz; this has a bedding or jointing, dipping 40° east. The same ridge also contains a coarse gneissoid rock, roughly schistose, and similar to the imperfect gneisses east of Darby. The schistosity strikes east-west, and the dip is 30° north; some of this gneiss is of fine grain.

GRANITE AND GNEISS.

It has already been stated that the range almost entirely consists of granitic and gneissoid rocks. The arrangement of these corresponds to the topographic development described above in this manner: The central mass is almost throughout a normal biotite-muscovite-granite; the regular eastern slope, even and continuous from Nez Perce Pass to Lolo Peak, is formed by a sheet of gneiss, dipping from 18° to 26° E. like that slope. The gneiss is a modification of the interior granite—in other words, is granite made schistose by pressure, and westward gradually changes into the massive granite. This gneiss is not only schistose, but also full of slipping planes, parallel to the schistosity, and on these slipping planes are commonly closely massed striations parallel to the dip of the gneiss. The structural importance of these remarkable facts will be discussed in a later paragraph (p. 47). In this place such detailed observations as are available will first be recorded.

North end of the range.—Going from north to south the gneiss is first observed at the mouth of Carlton Creek Canyon. It is here imperfectly schistose, dipping 15° to 18° E., and the belt is not wide.

Two miles farther up normal granite crops on the south side. As described on page 36, the gneiss borders the quartzite of the Lolo series, and the contact shows it to be of intrusive character.

A few miles farther south, due west of Florence, the gneiss has locally a steeper eastward dip of 30° to 35°, and fine cliffs of these rocks are seen, where the canyon walls cut the front plane of the range, 2,000 feet above the valley. Large white granite cliffs appear higher up in the canyon. The exposures in the lower part of Big Creek along the Curlew mine ditch are excellent; the gneiss dips 20° E., striking a few degrees east of north exactly parallel to the local trend of the foothills; the rock is medium-grained, dark gray, and very much sheared, the larger feldspars being pressed into lenticular shape and surrounded by films of fine-grained biotite. Striated slipping planes are very frequent, the striations being strictly parallel to the dip of the gneiss; large foils of secondary muscovite have in many places developed on the planes of schistosity.

At Bear Creek a small area of massive granite lies in the first foothills, but immediately west of this the normal gneiss begins and is well exposed with easterly dip of 18° at the entrance to the canyon, above the moraines.

Mill Creek section.—A good section is obtained along Mill Creek, the almost perpendicular bluffs of the lower canyon offering excellent opportunities for observation. The dip of the gneiss is from 15° to 25° E., probably averaging 18°. It seems a little steeper near the valley and flattens out slightly toward the west. Horizontally the width of the gneissoid zone is 1½ miles, corresponding to a thickness of about one-half mile. In petrographic character the gneiss is similar to that from all points described above. It is medium to coarse grained, consisting of elongated aggregates of gray feldspar and quartz separated and surrounded by streaks of black and white mica. Striated slipping planes are very common, as are also elongated pressed orthoclase crystals up to 1 inch in length. As the canyon is ascended the gneissoid structure becomes less apparent and about 2 miles above its mouth, where the whole bluff is beautifully exposed, the bottom is in granite and schistosity sets in by degrees toward the top. Much of the gneiss has the same grain as the granite; some layers are, however, finer grained and darker in color. This is clearly not due to any diversity of material, but only to the degree of crushing and compression. Some layers are so much sheared as to constitute a dark slaty mass traversed by innumerable wavy slipping planes. The petrographic examination, as well as that in the field, allows no doubt as to the original identity of granite and gneiss. From this point westward to the divide and far beyond it granite is the only rock present. The prevailing variety is medium to coarse grained, the average size of the grains being about 5 mm.; it contains much grayish quartz and whitish orthoclase or perthite and oligoclase, and besides small quantities of both biotite and muscovite in small foils, the former usually prevailing. Porphyritic orthoclase crystals up to 1½ inches in length are of frequent occurrence. The outcrops are brilliantly white; the rock weathers with a thin hard and white crust, fresher in appearance than the inside, the softer texture and slightly yellowish color of which becomes apparent when the surface is broken off the large glaciated outcrops. The granite in many places contains many streaks ("schlieren") and masses of darker, finer grained rock, with more biotite and sometimes rich in titanite. Usually they appear as if kneaded into the normal rock and exhibit a pseudo-schistose, contorted appearance probably due to fluidal structure. These masses are probably more basic, differentiated, and early cooled parts of the magma. Pegmatitic masses and dikes of all descriptions are conspicuously absent.

An east-west perpendicular jointing of the gneiss zone is a very common feature, producing the perpendicular walls of the canyons. In the granite the same structure is not so persistent. There is much jointing, to which the pinnacles of the ridges are largely due, but no striking regularity is exhibited. On glaciated outcrops an exfoliation is often noticeable, causing the surface to scale off in concentric, slightly curved slabs.

Like most of the canyons, Mill Creek bends south near its source and a small glacial lake, surrounded by a cirque of bare granite cliffs forms its head. On ascending this cliff, which is about 1,000 feet high, an excellent view is obtained over almost the whole extent of the Clearwater Mountains. The bare, glaciated outcrops seen in the maze of sharp peaks and serrated ridges northward as far as Big Creek and southward as far as Lost Horse justify the conclusion that the whole range is built up of granite. Westward the same kind of outcrops are seen on successive ridges as far as Grave Peak, which is a prominent granite point rising from a lower forested plateau. Rhodes Peak and at least parts of the Lolo Ridge appear to be of the same material. Similiarily, The Crags are clearly of granite. The whole gives a remarkable illustration of the extent of this rock.

Gneiss of usual structure and dip of 18° to 20° E. appears in the lower parts of Roaring Lion and Sawtooth creeks and numerous granite bowlders give evidence of the prevalence of this rock in the upper parts of the canyons.

Lost Horse section.—West of the heavy moraine piled up at the mouth of Lost Horse Canyon a high cliff of gneiss appears on the north side. The rock has the same structure and appearance as described above, but the dip is distinctly less. On the first cliff it is only 5° E. As usual many striated slipping planes lie parallel to the schistosity. On the south side, the dip is 15° and at the forks of the creek 10°. Smooth, glaciated outcrops begin 2 miles from the mouth. The section is different from that of Mill Creek inasmuch as the schistosity has a much smaller dip and probably in consequence of this extends much farther westward. At the big slide 11 miles from the mouth the rock is still a gneissoid granite, though the almost horizontal schistosity is much less distinct than at the mouth and sometimes is recognizable only with difficulty. Slipping planes occur very sparingly. The rock in the 1,500-foot bluff northeast of Twin Lakes, at the low divide (elevation 6,600 feet) still exhibits a slight but unmistakable schistosity dipping 8° W., instead of east. Almost the only macroscopical indications of the structure are the parallelism of the mica foils and the elliptical form of the large orthoclase crystals. The rock is otherwise identical with the Mill Creek type, only having a somewhat smaller amount of muscovite. Like that type, it weathers to brilliantly white crusts. It contains, also, a few dikes of slightly sheared pegmatite and some of the same ill-defined basic streaks as noted on Mill Creek. Parallel to the schistosity is a fairly well-defined jointing, the planes of which in places show striations parallel to the dip. A few less continuous joints, also with striations parallel to the dip, cross the first set, and dip about 20° E. There is, also, a well-defined east-west jointing, nearly perpendicular, or dipping 60° N. On these planes also occasional striations may be seen. A fine view is obtained westward down across the deep U-shaped depression of Moose Creek, dark with thick lodgepole-pine forest and surrounded by steep cliffs of white, granitic rock.

Rock Creek section.—The next canyon south is that of Rock Creek, resembling the others in its straight east-west course and deeply incised valley. Cutting into a more elevated part of the range, the slopes in its central part are exceptionally high in some places, being as much as 4,000 feet. While the main fork has a fairly even grade, the south fork heading in the glacial cirques below El Capitan descends over precipitous cliffs to the junction. In the first cliffs north of Como Lake the gneiss is very schistose, but in varying degree, some belts being reduced to the appearances of a mica-schist, while others still show clearly the porphyritic orthoclase crystals. The dip is 22° E. Local bends in the schistosity were observed changing the dip from steep to flat and back again to the normal inclination. The heavy falls of the South Fork descend over gneiss with flat easterly dip. On the main creek the gneiss continues with dip of 15° E. for 2 miles above the lake. The moraines at the mouth of the canyon contain abundant granite bowlders. Gneiss was again noted in the lower part of Tincup Canyon and has here a dip of 27° E. The observations on Rock Creek and Tincup Creek were mainly by Mr. G. W. Stose. Neither El Capitan nor the Como peaks were ascended. From their appearance it is likely that they consist of gneissoid rocks with flat dips.

Trapper Creek section.—A few miles south of the forks of the Bitterroot, Trapper Creek empties into the South Fork after a short course of about 6 miles, the eastern slope of the range narrowing considerably south of Tincup Creek. Trapper Creek heads among the cirques of the peak of the same name, the highest elevation in the range (Pl. IX, *A*), and a heavy moraine has accumulated in front of its canyon. Near its mouth the prevailing rock is the granite of the eastern foothills (p. 32), cut by frequent dikes of granite-porphyry. This continues for one-fourth of a mile above the mouth. The moraines, containing some gneiss but chiefly granite bowlders, cover the underlying formation; 3 miles up from the main river the first high bluffs on the north side consist of gneiss, partly normal with the characteristic drawn-out feldspar crystals, partly more crushed and chloritic; a few slipping planes with usual direction of striations were observed. The dip averages 15°, but varies in detail from 10° to 40°. One mile farther up there is no gneiss in the lower outcrops, but only a coarse granite, while the débris fallen from the top of the bluff still consists of gneiss. The granite shows a few joint planes dipping 20° E. From points well up on the slope fine views are obtained up toward the sharp peaks and cirques at the head of the creek, all apparently granite, while the front slope of the range, beautifully shown by its line of intersection with the south side of the canyon trench, clearly consists of a sheet of gneiss gradually carried up to points with elevations of about 8,000 feet. Beyond, the cliffs break off westward and granite prevails.

South end of the range.—South of Trapper Creek the strike of the gneiss sheet swings to north-northeast, the dip remaining the same or about 15° (fig. 3). The even slope of the frontal plane is carried up at this angle to the very divide, the descent to the Bitterroot River being here less than 6 miles. The deeply incised Boulder Canyon is passed 8 miles above Trapper Creek; in a few miles more the West or Nez Perce Fork is reached. This valley opens up wide and broad in contrast to the narrow canyons of the main river, or South Fork, above the junction, and from high points in this vicinity a most interesting view is obtained, showing the manner of termination of the range. At the junction of the forks and for several miles along the West Fork, which is followed by the Nez Perce trail, massive granite prevails, penetrated by many dikes of granite-porphyry.

From the vicinity of Boulder Creek the front plane of the range begins to swing in a still more easterly direction, but is continued with the same general dip and forms the whole northern slope of the West Fork Valley.

The first creek coming down this slope about 2 miles above the junction of the rivers was ascended by Mr. Stose, who found the first outcrops about 800 feet above the valley. Here the normal Bitterroot gneiss is encountered with a strike of N. 52° E., and a dip of 40° SE. Besides this, there is a rough bedding dipping 8° SW., and joint planes with evidence of slipping, which strike N. 75° E. and dip steeply to the north.

Beyond this point the frontal plane is seen to continue for several miles farther west, turning gradually to an almost east-west strike. Gneiss appears to occupy this whole slope to the summit or main divide and the dip of the schists is clearly seen to conform to the slope of the plane. Toward the head of the West Fork it appears as if this gneiss gradually merged into the prevalent granitic mass. This is the conspicuous southern termination of the Bitterroot Range. South of the West Fork geology and topography undergo a complete change.

STRUCTURE AND PHYSIOGRAPHY.

The topographic description of the range has shown the existence of a most persistent plane dipping 15° to 26° E., which forms the slope inclined toward the valley, and which continues with due north-south direction from Lolo Peak to Boulder Creek, where at the southern end of the range it gradually swings westward, retaining the same dip, now turned southerly. A short distance west of the bend the plane appears to die out. It can not have been produced by erosion, and the only alternative is to consider it as a structural plane. An examination of the geology reveals the fact that this plane is formed by a layer of gneiss, which petrographic methods show to have been derived by compressive or shearing stress from the normal massive granite forming the mass of the range. The gneiss zone conforms in strike and dip with the structural plane, and moreover shows persistently and uniformly, by little slickensides and striations, evidence of a movement parallel to the dip of the gneiss and the structural plane. The aggregate movement along all these slipping planes may well have been very large. While the gneiss zone usually is thin, and ordinarily less than 3,000 feet in width perpendicularly to the strike plane, it flattens in the Lost Horse section, and a roughly gneissoid granite with nearly horizontal schistosity extends westward at this place as far as the main divide. The strongest schistosity and the most closely massed slipping planes always occur farthest east along the structural fronting plane.

The only conclusion which can be drawn from these facts, remarkable and unique as far as my knowledge goes, is that a uniform dislocation of great extent has taken place all along the east slope of the range. The main dislocation was accompanied partly by direct movement on minute and numberless fault planes (slipping planes) and partly by molecular movements long enough sustained to produce a typical schistose structure within the affected zone. The unusual feature in this dislocation is the intimate combination of actual fault planes with schistose structure, showing in the clearest manner, to my mind, the transitions between schistosity and jointing, and between molecular and molar movement, and indicating that both result from the same force. Taking all the data into consideration, it does not seem as if the depth at which this schistose structure was produced could have been more than a few thousand feet at the utmost. Any attempt to place the formation of this schistosity at great depths from the surface must lead to wholly improbable depths of erosion and amplitude of faulting movement. Another unusual feature consists in the flat dip of the fault plane; this flat dip is no doubt the cause of the remarkable topographic preservation of the fault plane.

The minimum of actual movement on the fault plane is clearly measured by the width of the gneiss zone along this plane from the summits to the base of the mountains, a distance of 2 or 3 miles. The vertical component of this movement would be 4,000 to 6,000 feet, and this again must be considered as a minimum.

Admitted that we have here a flat fault of great extent, the next question is as to the absolute direction of the movement. That the relative movement took place parallel to the dip of the gneiss is clear beyond question from the observed facts. It was first thought that the deformation of feldspar crystals in the gneiss would elucidate this problem. No satisfactory answer was obtained in this way, though more detailed studies perhaps would be productive of better results. The deformation of the crystals seemed to have proceeded under such heavy pressure that a flow rather than a break resulted, and it was impossible to tell whether the hanging wall had relatively moved up or down. The slickensided slipping planes, on the other hand, never break across a crystal, but follow in gentle curves the aggregate of recrystallized mica scales, on which, of course, the movement was most easy.

Approached from a different direction some elucidation of this problem might possibly be attained. Throughout the Salmon River Mountains, as well as in the Clearwater and the Coeur d'Alene mountains, the configuration of the ridge lines shows that they may be considered as the remnants of a gently undulating plateau raised by epeirogenic forces and dissected by active erosion. The average elevation of this wide plateau of central Idaho is perhaps 7,000 feet. It extends far north and south of the Bitterroot Mountains, which stand as a boldly raised block 1,000 to 3,000 feet above this level. It seems unavoidable to connect this uplift with the fault at the base of the range, and thus we arrive at the conclusion that the foot-wall side of the fault has been raised relatively to the hanging side; in other words, the fault is a normal one. Considering again that the Bitterroot Valley is an extremely well-marked and probably very deep depression almost coextensive with the range, and that borings would probably show it to be a structural depression without outlet, we come again to the conclusion that the hanging has subsided relatively to the foot of the fault; in other words the fault is probably a normal one.

There are interesting facts which prove that this subsidence has taken place and that it has not even yet ceased. At the Curlew mine, located in the first foothills in the northern part of the range (p. 86), mining operations have disclosed a fault dipping 45° E., along which a body of galena ores was found and mined. I am reliably informed that this fault separates the lime and quartzite of the foot wall from a hanging wall of gravels with logs of carbonized wood. A heavy clay seam lies along the fault. This fault is exposed to a vertical depth of 500 feet. As stated, the first foothills here consist of limestone and quartzite of uncertain age; these rocks have not been rendered schistose like the granite, but are thoroughly crushed. The hill back of the mine is cut by slipping planes, and a little higher up the narrow zone of limestone and quartzite borders along a fault contact toward the normal gneissoid zone. Along this contact, far removed from mining operations, a fault took place in 1898, the downthrow being on the eastern side; the throw is still visible along a distance of 1,500 feet north and south, and amounted to 1 or 2 feet.

Still another line of evidence points in the same direction. As described above, (p. 40) many straight canyons furrow the east slope of the Bitterroot Mountains. They have a very steep inclination, but below the upper zone of glaciation in which falls over granite ledges are frequent, there is no reason why they should not have average regular gradients. While this holds good in most cases, there are in some places curious anomalies. Where Mill Creek emerges from the range, for instance, the first mile of canyon has a grade of 400 feet in 1 mile, while the regular grade is 200 feet.

For several miles south of the mouth of Carlton Creek Canyon, also in the northern part of the Bitterroots, a distinct escarpment perhaps a few hundred feet high, runs along the foot of the range. An exceedingly heavy fall of 600 feet in 1 mile marks the mouth of the canyon, the lower part of this distance being over morainal bowlder heaps, the upper part over gneissoid bed rock.

It seems thus that during recent or even during the present time, short and irregular displacements occur along the foot of the range and in all cases the downthrow is on the valley side.

From all that precedes it is plain that the east front of the Bitterroot Range is determined by an enormous flat fault. It further follows with great probability, that the fault is of normal character, and that the dislocation is a double one, involving at the same time downward movement of the hanging and upward movement of the foot. On the west side of the Bitterroot Mountains the upward movement of the block seems to have been so distributed as to leave few indications of its presence. Such double movements, involving a simultaneous raising of the heaved block and sinking of the thrown one, are probably not uncommon. Studies along the east slope of the Sierra Nevada have led me to believe that the movement along this break has been of this character.[a] The flat dip of the fault plane indicates

---

[a] Jour. Geol., vol. 4., 1896, p. 899.

pretty clearly that gravity does not alone suffice to explain this movement, for as pointed out by Mr. G. F. Becker,[a] friction would overcome gravity on a slope of the fault plane of 45°.

In presenting these conclusions I am not unmindful of the cursory character of the field work and fully appreciate that more detailed examinations will throw a much more vivid light on the extremely interesting questions of mountain building and rock structure involved. The main facts seem so plain, however, that they scarcely admit of any other interpretation than that advanced. It is believed that the Bitterroot Range will always remain one of the most important localities in the world for the study of the development of schistosity and distributed faults.

It seems unavoidable to recognize the intimate and inseparable relation between schistosity and faulting, or, in other words, between molecular and molar rock movements, in the sheared zone of the Bitterroot Range. The striated slipping planes of the granite-schist are often very closely massed, as many as twelve of them sometimes occurring in a thickness of an inch. Their surfaces are covered with micaceous aggregates or large, new-formed flakes of muscovite, and they curve gently about the lenticular, pressed aggregates of former larger feldspar or quartz grains. The latter are sometimes broken by a series of small fissures inclined about 45° to the plane of schistosity and filled with secondary quartz. (Pl. V, A, B.) Many of the slipping planes form blind joints, nonpersistent or gradually fading.

The two differing views of the relation of schistosity and jointing may be briefly recalled: Mr. Van Hise holds[b] that the two are distinct phenomena. In the upper or "zone of fracture" of the earth's crust, bedding fractures, faults, joints, fissility, etc., are formed almost wholly by the breaking of individual minerals and by differential movements between them. Rock flowage, producing schistosity, mainly occurs in a very deep-seated zone by innumerable fractures of the mineral particles, or by recrystallization; in brief, it is a molecular movement in which mass fractures play a subordinate part.

On the other hand, Mr. G. F. Becker[c] holds that there is no essential difference between jointing, cleavage, and schistosity, all of those phenomena being due to weakening of the cohesive force on planes of maximum slide or maximum tangential strain. Accordingly, schistosity and jointing, molecular and molar movement, may both be produced at the same time, and molecular movement is not necessarily confined to great depth. The observations in the Bitterroot Range seem to confirm Mr. Becker's views.

---

[a] Bull. Geol. Soc. Am., vol. 4, 1893, p. 47.

[b] Van Hise, C. R., Principles of North American pre-Cambrian geology, with an appendix on flow and fracture of rocks as related to structure, by L. M. Hoskins: Sixteenth Ann. Rept. U. S. Geol. Survey, pt. 1, 1896, p. 589.

[c] Becker, G. F., Finite homogeneous strain, flow and rupture of rocks: Bull. Geol. Soc. Am., vol. 4, 1893, pp. 13–90.

If we should tentatively endeavor to outline the history of the Bitterroot Range it would perhaps be something like this:

During the Cretaceous or earliest part of the Tertiary period the surface of the high plateau of the Clearwater Mountains extended across this whole region and included a large part of western Montana. On this uplifted plateau a drainage system had been outlined, and of such origin, we may assume, was the Bitterroot River. The divide toward the Clearwater drainage was situated along the present summit of the range about 10 miles east of the present position of the same divide. The elevation of this divide was 2,000 to 3,000 feet lower than at present.

Now followed the great break along the line of the Bitterroot River. A fault plane was formed with flat easterly dip and a movement of probably at least 20,000 feet along the dip of that plane took place. This movement had the twofold effect of raising the Bitterroot Range to its present position on the west and of deepening the depression of the Bitterroot Valley.

The gently sloping structural front plane of the Bitterroot Range being thus established a new constructional drainage was at once laid out on the even surface and naturally the creeks would be numerous and straight, just as we find them to-day similar in direction but more deeply incised. Grade and precipitation combined to increase their corrading activity; soon they were robbing the streams of the then broad plateau on the Clearwater side and the divide migrated westward to its present position with the gradual capture of new territory. The general bending toward the south, common at the head of the Bitterroot streams, is no doubt due to that same prevailing tendency in the headwaters of the old captured Clearwater streams.

Finally, as the last important stage, the névé fields increased in size and soon a continuous ice sheet covered the summit region. During the Glacial epoch the summits were carved out to the bewildering mass of craggy arêtes separated by precipitous cirques which now crown these mountains.

GLACIATION.

There is excellent evidence that during the Glacial epoch the Bitterroot Range along its whole extent was covered by a continuous sheet of névé, from which glacier tongues extended down into every important valley. It forms thus one of the very largest glaciated areas known in the Cordilleran region of the United States, and only rivaled in extent by the glaciers of the Sierra Nevada and the Cascade Range. There is no evidence of glaciation along the Lolo Fork of the Bitterroot River; nor is it probable that any but local foci of glaciation existed from Lolo Fork northward along the Coeur d'Alene Mountains. Similarly the

Nez Perce Fork, or West Fork, at the southern terminus of the range, is free from glaciation, and no evidence of anything but very local accumulations of ice were noted upon the headwaters of the South Fork south of Nez Perce Pass.

The U-shaped form of the valleys, the moraines accumulated in front of their canyons and within their troughs, the polished outcrops of granite throughout their extent, all bear undoubted testimony to the existence of this ice sheet. The main canyons as well as the tributary gulches terminate everywhere in glacial cirques. In short, all the familiar phenomena of extensive glaciated areas are represented in the Bitterroot Range on a grand, instructive scale. Only the highest peaks and ridges appear to have projected above the general level of ice and sand and between the projecting glacier tongues. The foothills of the range also rose to an elevation of 2,000 feet above the valley below the line of permanent snows.

Not in all cases did the glaciers extend down to the mouth of the canyons where they opened on the valley. There is clearly evidence of increasing strength of the glacial action toward the southern end of the range. Here large morainal masses several square miles in extent and with ridges sometimes a thousand feet in height lie in front of the foothills.

Beginning from the north, the Carlton Canyon was first examined. The canyon pass is narrow and V-shaped, and it is doubtful whether any of the material in front of it is of true glacial origin. Heavy masses of bowlders lie on both sides of the canyon at its mouth and extend in fan shape for a distance of half a mile. Possibly these accumulations are the result of local cloud-bursts or torrential action.

At the mouth of Big Creek Canyon the gravel terraces extend up to the foothills of gneiss, and there are no accumulations of bowlders in front of the canyon. Its lower part is V-shaped and sharply incised without polished outcrops or other indications of glaciation.

At the mouth of Mill Creek Canyon a sloping plain of gravel and sand extends down to the river 400 feet lower. Over this gravelly slope Mill Creek flows in a shallow cut at most 30 feet deep.

The first signs of glacial action are seen at Sears's ranch, near the foothills, and consist of a big pile of granite bowlders, some of them 8 feet in diameter. A very heavy moraine lies on the south side of the creek near the same place and close to the mouth of the canyon. The surface forms of these bowlder masses indicate unquestionably their morainal origin. Little ridges and shallow ponds abound. The height of this moraine is several hundred feet, and on its southern side no rock in place appears for a distance of half a mile farther west, while on the northern side outcrops begin much lower down. Near the mouth the canyon presents no trace of actual ice erosion either in the wearing or rounding of the outcrops, so that it is not probable that the ice stream itself reached this point.

Like all the Bitterroot canyons, the grade of Mill Creek is very heavy. On the gravel planes in front of the range it is 200 feet to the mile. At the mouth of the canyon for a distance of about 1 mile it is much heavier, amounting to 400 feet in that distance. Above this the grade is somewhat lighter, until several miles farther up an abrupt fall of over 200 feet occurs below the place known as "the great slide."

Above the great slide the grade is again fairly moderate up to the lake, and amounts to approximately 200 feet per mile. Morainal matter fills the comparatively narrow bottom of the canyon for a distance of 1 or 2 miles above the mouth. At about this distance the valley becomes much more U-shaped, and frequently smooth and flat rocks occupy the creek bottom. The width of the valley increases westward, the bottom is filled by morainal material and supports a thick growth of lodgepole pine besides rising more steeply and frequently showing polished surfaces of granite. An interesting feature is the nearly perpendicular cliff a thousand feet in height which follows the north side of Mill Creek for 3 miles above its mouth. Pls. VI, B, and VII, A, illustrate the topographic features of these and the shape of the canyon. Great slides of angular granite blocks descend into the canyon at many places, and sometimes on the forested slopes are long clearings illustrating the devastating power of snowslides.

At the head of the canyon a small lake dammed by morainal material occupies a large glacial cirque surrounded by polished slopes of granite. Even the saddle on the divide a thousand feet above the lake shows evidence of glacial polish. It seems that the névé fields covered everything along the summit region except the highest peaks.

Blodgett, Sawtooth, and Roaring Lion creeks are almost exact counterparts of Mill Creek. The remarkably straight and sharp ridge between Sawtooth and Roaring Lion is illustrated on Pl. VII, A, as viewed from Wards Peak. In all cases the moraines accumulated at the lower end of the canyons are comparatively insignificant.

South of Wards Peak the conditions change. Lost Horse, Rock, Tincup, and Chaffin creeks are all distinguished by great development of lateral and terminal moraines, the extent of which are illustrated by the panoramic view on Pl. X, taken by Mr. G. W. Stose from the terrace northeast of Darby on the east side of the river. The plate shows the bottom lands and low terraces along the river, above which rise the grassy and open high terraces; these again lean against broad and rounded morainal hills, heavily forested, as a rule. Back of these rise the jagged peaks and ridges of granite and gneiss. Pl. I shows the areal extent of the heavy moraines of the south end of the range. At the mouth of Lost Horse Canyon moraines occupy several square miles, the most prominent one forming an 800-foot ridge on the south side, reaching up to an elevation of 5,000 feet. The creek has cut through the terminal moraine, and gravel terraces skirt the east front of the moraines 200 feet above the river. The canyons of the forks of Lost Horse Creek are practically duplications of those of Mill Creek, the bottoms being U-shaped and often showing extensive and smooth surfaces of ice-worn granite, and elsewhere occupied by a thick growth of lodgepole pine. The sides are precipitous cliffs with many enormous rock slides. Near the head of the Middle Fork lies a heavy bottom moraine which has dammed the Twin Lakes at Lost Horse Pass.

In a similar manner, high lateral moraines have been piled up on both sides of Rock Creek. The terminal moraine has been partly cut through by the stream, but above it still lies the beautiful sheet of water dammed by it, which is called Como Lake. From its pine-clad shores a magnificent view is obtained up toward the jagged summits of Como Peak, 5,000 feet above the lake. (Pl. VIII, B.) At the mouth of Tincup Canyon similar though smaller moraines have accumulated. The ice stream was deflected southward, so that the lateral moraine on the north side swings around in a graceful curve.

From this point southward the distance to the main divide is less, and the glaciers were correspondingly smaller.

The moraines of Trapper Creek form a fine semicircle, opening southward. They are 3 miles long and 300 feet high, and extend almost to the Bitterroot River. The last large tributary from the western slope of the Bitterroots is Boulder Creek, the moraines of which project to the river and have forced it to cut a cliff in the eastern foothills.

The 200-foot gravel terrace skirts the moraines down to the junction of the two main forks of the river and continues for some distance up the East Fork, while the South Fork south of Chaffin Creek enters a narrow canyon.

From all appearances, the glaciation of this range extended to a very recent date. While in all cases, excepting Rock Creek, the principal creeks have trenched the terminal moraines, but little evidence of erosion is to be seen beyond this cutting, and in many places higher up these torrential streams still flow over broad and flat glaciated outcrops, not yet having had time to excavate new canyons in them. The surface of the moraines is often remarkably smooth, due no doubt to the sandy and crumbling character of the granite, and is generally covered by a fine growth of yellow pine.

On the western side of the range the glaciers must have been still more extensive than on the slope facing east. The canyon system is here more branching; the precipitation was also probably heavier. The result would be that several ice streams would meet, and, coalescing, extend far down into the Clearwater drainage. Just how far west they reached has not been ascertained, but from the available information as to the size of the glacial cirques at the summit and the conditions of the rivers farther down it is probable that along the Lochsa, Moose Creek, and Selway forks this distance was about 20 miles from the main divide. Very likely the névé fields of the northern part of the Bitterroots extended across to Grave Peak, 15 miles distant from the main divide.

SOUTH FORK OF BITTERROOT RIVER.

TOPOGRAPHY.

At the junction of the East and South forks of Bitterroot River the valley is still an open one, with bottom lands 1 mile wide and broad gravel terraces, the latter continuing for several miles up along the easterly branch of the stream. Broad and partly timbered ridges ascend with comparatively easy slope from the East Fork, their summits forming a nearly level sky-line 2,000 or 3,000 feet above the valley. A low granite ridge projects from the Bitterroot foothills, connecting with the mountains surrounding the East Fork, and across this ridge the South Fork breaks in a narrow canyon, suggesting that a former valley, nearer to the Bitterroot Range, has been filled by morainal débris and a new trench excavated through the granite ridge in post-Glacial time. For a distance of 15 miles above this canyon the South Fork flows in a narrow valley which, however, opens up a little immediately south of the canyon through the granitic ridge mentioned. The bottom, though narrow, is flat, and occasionally fragments of a low terrace appear above the stream. Where the Nez Perce trail leaves the South Fork the West Fork Valley branches and forms a wide, open depression. Above the West Fork the main branch continues narrow, with flat bottom and sides rising steeply for a few hundred feet up to a sloping bench, from which heavily forested broad ridges run up to a similar main divide very different from the narrow arêtes and pinnacles of the Bitterroot ridges. These conditions continue for several miles above the West Fork to near Hughes Creek (elevation, 5,000 feet), where the valley opens to a width of half a mile and is surrounded by bluffs of rock 1,000 feet high. Above this creek the Bitterroot River splits in many branches and the rounded forest-covered ridges of the Salmon River divide are soon reached. The highest elevation of this is found at Blue Nose Mountain, which attains 8,300 feet, but the average height of the dividing ridge is considerably less.

GEOLOGY.

GRANITE.

For a distance of 25 miles above the main junction of the rivers, or up to the vicinity of Blue Joint Creek, granitic rocks prevail in the canyon of the South Fork. This granite is, on the whole, similar to that described from the eastern foothills of the Bitterroot Valley. Its contact with the southern quartzite is not well exposed, and nothing definite can thus be said about the relations of the two rocks. The granite is rich in dikes of granite-porphyry, sometimes appearing very similar to rhyolite.

In the first canyon above the main junction a coarse biotite-granite prevails, with a few inclusions of gneissoid rock. At the mouth of Trapper Creek the granite contains many dikes or irregular masses of reddish granite-porphyry. Three miles above this creek a coarse granite with very large feldspar crystals was noted, and again opposite Boulder Creek are cliffs of the same reddish rock. Similar coarse, reddish biotite-granite, rich in orthoclase, appears near Rumbo Creek, here containing many dikes of granite-porphyry, a rock which is especially abundant in the gravels of this creek. Large masses of alternating granite-porphyry and granite outcrop for 2 miles above this point.

Mr. Stose ascended the West Fork on the Nez Perce trail for several miles and reported first granite with streaks and inclusions of gneissoid rocks similar to the exposures east of Darby and farther up much granite-porphyry, together with pegmatitic rocks and some contorted gneiss. The sheared granite of the long slope from the Bitterroot Mountains, north of the West Fork, has already been referred to (p. 46).

SEDIMENTARY SERIES.

An extensive series of sedimentary rock of doubtful age is encountered on the upper South Fork of the Bitterroot. It begins near the mouth of Blue Joint Creek, a tributary with several branches coming down from the west about 10 miles above the point where the Nez Perce trail leaves the valley of the South Fork. White and heavy-bedded quartzites prevail, the strike being northwest, the dip 80° S. A long ridge leads from Blue Joint Creek up to the high points south of Nez Perce Pass, called Crown Peak or Castle Rock, and Bare Cone; going up, the quartzite is soon replaced by a belt of granite with many dikes of granite-porphyry. Then quartzite containing many dikes of granite-porphyry begins again and occupies the high ridges of Bare Cone, having an elevation of about 8,000 feet. The summit of Crown Peak is formed by a neck of rhyolite.

Above Blue Joint Creek the same heavy-bedded quartzite continues up to near the head of the river at Mineral Point, and its precipitous bluffs follow the stream for several miles. The same rock occupies almost the whole of the Hughes Creek drainage, and probably also reaches up to the Clearwater divide, 8 miles distant toward the east.

At Mineral Point Meadows are quartzitic schists with varying dip, generally about 30°; the dip is northward at the lower end of the meadows, changing to east or south near Copper Queen mine. The schist is here more micaceous and fissile, but still distinctly sedimentary. Going upstream the schists become still more micaceous. A little coarsely crystalline limestone is met on the road along Beaver Creek one-fourth mile above its mouth. Many dikes of granite-porphyry and diorite-porphyry appear here. One mile above the mouth on the same creek the schists change to typical gneiss. The contact is apparently not of intrusive character, but the two kinds of rock rather appear as if welded together. Probably the quartzitic schists rested on gneiss, and the exact contact line has become blurred by action of metamorphic processes As stated above, there is nothing to prove the age of the sediments, except the probability that they once rested on a gneiss which has the characteristics of an Archean series. The similar conditions might justify a guess that the quartzites are of the same age as the Lolo series, possibly pre-Cambrian. The slates and quartzite are known to extend for many miles eastward, or at least as far as Gibbonsville, 12 miles distant from Bitterroot River, where Mr. W. H. Weed observed a series of steel-gray quartzitic schists dipping 35° to 65° W.

TERTIARY VOLCANIC ROCKS.

Although some of the dikes in the granite for the first 20 miles south of the main fork of the Bitterroot River are somewhat rhyolitic in aspect, no areas of this rock are encountered until near the mouth of Blue Joint Creek. At this point a flow of normal rhyolite filled the valley to a depth of several hundred feet and is now eroded to high bluffs showing on both sides. A small mass of rhyolite forms the summit of Crown Peak or Castle Rock. Areas of this rock are also found on Slate Creek, on the west side of the valley, and south of Blue Joint Creek this rhyolite may even reach the Clearwater divide. On Coal Creek, coming in from the west, 2 miles above Blue Joint, Mr. Stose found the rhyolite extending up to at least 1,000 feet above the valley. In its uppermost part it contains thin interbedded clays with seams of lignite (see p. 112).

The next exposure of volcanic rock is found 10 miles farther up, on Lick Creek, on the east side of Mineral Point mining district. These thick flows of rhyolite, which extend at least 1,000 feet above the valley, consist of several flows, some glassy, others felsitic. In part they are interstratified with tuffs or sedimentary material resembling lake beds, the latter containing traces of lignite and some

Unio shells. According to Mr. W. H. Weed much rhyolite is also seen along the East Fork of Bitterroot River.

While there is no direct proof of the age of the rhyolite, it is probable that, like nearly all of that rock in Montana, it is of middle Tertiary or Miocene age. Its topographic extent is perhaps the most interesting feature, as it clearly proves that the configuration, depth of canyon, etc., in the upper South Fork were about the same as at the present time. As far as this fork is concerned, subsequent events have consisted rather of repeated fillings and retrenchings of the pre-Miocene excavations than of any important deepening of the channel.

QUATERNARY.

The Quaternary deposits along the South Fork of the Bitterroot are limited to the bottom lands and terraces along the river. Though the bottom lands are narrow and of slight extent, they persistently follow the river, indicating a recent damming, which deprived the river of its full erosive power. In few places the water flows over exposed bed rock. Above these alluvial areas are several terraces, preserved in fragments, indicating that at a comparatively recent date gravels filled the canyon to a depth of over 500 feet. The analogy with Lolo Fork (p. 38) is apparent. In the lower South Fork one direct cause for the accumulation of gravels above is found in the fact that during the Glacial epoch the moraines of Trapper and Boulder creeks produced effectual dams of considerable height. Whether this is sufficient or whether the uniform filling of Lolo, the Bitterroot Valley, and the South Fork to a height of 400 to 500 feet above the present water level was due to more universal causes acting farther down on the Clark Fork of the Columbia can not be regarded as fully settled. The erosion which followed the maximum of deposition was clearly of active character and is responsible for the steep bluffs which now in many cases skirt the river.

As stated above, three terraces, 30, 100, and 200 feet above the river, form the broad valley at the junction of the two forks. Immediately above this the South Fork emerges from a narrow canyon extending for a few miles and then opening up a little at intervals. A low terrace, 20 feet above the river, follows closely, and in places are remnants of another 200 feet above the stream. At the mouth of West Fork the lower terrace is well developed, and another of gravel and silt appears on the west side 200 feet above the river. Near the mouth of Blue Joint Creek gravel was noted from 500 to 550 feet above the river. At the mouth of Hughes Creek gravel bars are found at 20, 50, 200, and 500 feet above the river, thus reaching up to elevations of 5,660 feet.

CLEARWATER MOUNTAINS.

TOPOGRAPHY.

As defined above (p. 13) the Clearwater Mountains form a roughly rectangular area 70 miles wide and 100 miles long (from north to south) bounded on the south by Salmon River, on the north by the North Fork of the Clearwater, on the east by the Bitterroot Range, and on the west by the Columbia lava plateau which begins a short distance east of the South Fork and the main Clearwater River. This broad area of about 7,000 square miles can not be regarded as a mountain range due to orographic forces acting along certain lines, but must be considered as due to a continental uplift acting evenly over the whole area. It is, briefly, an elevated plateau, of approximately uniform height, now so deeply dissected by such an intricate system of canyons that at first glance it has almost lost its once doubtless prominent plateau features.

RIDGES.

Three primary divides extend westward across the whole width of the Clearwater Mountains from the backbone of the Bitterroot Range. The most northerly is the ridge between the North Fork and the Lochsa Fork. Immediately west of the broad, low depression of Lolo Pass this divide rises to elevations of from 6,000 to 7,000 feet, and its bleak summits of white granite with scattered dark groups of firs extend for 50 miles west-southwest, maintaining about the same height. From near Rocky Ridge it slopes gently with many low spurs, within a distance of 18 miles, down to the basalt plateaus of Weippe, which are situated about 3,000 feet above sea level.

The Lochsa-Selway divide, of about the same length, is more broken and irregular; and its peaks are also higher. From the Bitterroot Range it forms a succession of points and high ridges, such as Grave Peak (elevation 8,287 feet), Diablo Peak (elevation 7,531 feet), and near the western end of the ridge, The Crags (elevation 7,923 feet). Ten miles west from the bare granite points of the Crags, the thickly forested ridge descends abruptly to the junction of Lochsa and Selway forks. Toward its head the Selway Fork splits in a great number of branches, separated by complicated ridge systems. Many of these head toward the Salmon River divide, which is a high ridge with many and sharp curves due to tributaries cutting into it from north and south.

The Salmon River divide begins a few miles northwest of Blue Nose Peak (at the head of the South Fork of the Bitterroot) and continues westward for fully 80 miles as a high, broad ridge, sparsely timbered and sometimes with swampy meadows surmounted by a few points exceeding 8,000 feet in elevation. Above the general level rises a bare reef of rocks extending for 12 miles north and south and marked by the three points of Caseknife Mountain, Salmon Peak (elevation 8,881 feet), and McGruder Mountain. This reef is sculptured by glacial erosion and the gulches leading up to it end in rocky ampitheaters. From the slopes magnificent views are obtained of the western Clearwater region and the Salmon River Canyon and the mountains south of it. This is illustrated by fig. 4, reproduced from a photograph by Mr. J. B. Lippincott, which shows well the plateau character of the broad forested but fire-scarred ridges west of Salmon Mountain continued south of the 4,500-foot gash of the canyon by an equally well-marked general surface level. From the main divide long backbones reach out toward

Fig. 4.—View south across Salmon River Canyon, from south slope of Caseknife Mountain, showing plateau character of Salmon River Mountains.

the canyon and suddenly break off in precipitous descents. This sort of country continues for 40 miles west of Salmon Mountain to Buffalo Hump. The lonely trail follows the ridge for 24 miles in a bee line, but much more than that, counting its windings through fir-covered saddles and over bare rocky ridges. It crosses Little Salmon Creek near its head, the swampy meadows surrounded by thick timber having an elevation of only 5,200 feet, but ascends immediately west of this creek to the plateau elevations of over 7,000 feet. At intervals views are obtained north and east over the Selway drainage—a labyrinth of forested ridges sloping gradually at first toward the separating canyons and then dropping off to the streams in precipitous descents. Against the skyline are sometimes visible the snowy crests of the Bitterroot Mountains, among which towers the isolated El Capitan.

West of Little Salmon Creek the ridge branches. The divide between the South Fork and the Selway continues with the same characteristics west-northwest. The mass of Bilk Mountain and Indian Park projects northward, overlooking the Selway Canyon, while the main part of this ridge extends north of Elk City and culminates in Anderson Peak, Bald Mountain, and Pilot Knob, all about 7,000 feet high. From the two latter points the plateau of the Clearwater Mountains descends abruptly to that of the Columbia River lava, 4,000 feet lower.

The view from Bald Mountain is instructive and impressive—even magnificent—but, unfortunately, not easily shown in a photograph. Eastward lie the Clearwater Mountains, with comparatively level crests and a maze of canyons and ridges, a wild and lonely country across which the trail leads for 115 miles, the first 80 of which does not disclose a settlement or even a miner's cabin, while westward, and 4,000 feet below, spreads the undulating plateau of Camas and Kamiah prairies, checkered with waving wheat fields or wild grass, yellow or golden brown at this autumn time. Rarely are wilderness and culture brought into so close contact upon such a tremendous scale. Nothing but this high lava plain is visible as far as the eye reaches toward the west, far into the State of Washington.

The southerly ridge, forming the divide between the South Fork and the Salmon River, continues 35 miles west of Little Salmon Creek with the same general aspect of broad forested ridges, with long and gentle slopes toward the north and abrupt descent southward toward the great canyon of the Salmon; then the high isolated complex of Buffalo Hump rises above it to 8,800 feet in elevation. West of Buffalo Hump the ridge divides again. One branch bends northward and drops off suddenly to Camas Prairie at Mount Idaho, while the other one continues 30 miles west, sagging to a gentle plateau of 6,000 feet elevation at Florence, rising again in the great bend of Salmon River to 8,000 feet, and then suddenly dropping off 6,000 feet into the canyon of that river. From this most westerly point of the Clearwater Mountains, across the deep chasm of the Salmon and Snake canyons, the skyline is formed by the lava plateau, which here is higher, reaching 5,000 or even 6,000 feet in elevation. To the south, across the Salmon River, rise the first spurs of the Seven Devils, an outlying group of sharp peaks, surrounded on the south and west by the Columbia River lava, but connecting on the east with the Salmon River Mountains.

CANYONS.

The widely branching forks of the Clearwater River, which empties into the Snake at Lewiston, drain a territory of nearly 8,000 square miles. The streams flow throughout in deeply incised canyons, usually so narrow and rocky as to rarely afford space for a trail along the water line. Only along its lowest course from Stuart to Lewiston is the Clearwater lined at intervals with small strips of bottom lands. For 60 or 70 miles above Lewiston it has cut a trench 1,000 to 2,000 feet deep, with precipitous slopes, through the horizontal beds of the Columbia River lava; beyond this limit begin the Clearwater Mountains, and in this province the canyons are less precipitous but far deeper, 4,000 feet being a common vertical distance between the summit of the ridges and the bed of the stream. The slopes are fairly gentle for some distance from the summits, but become much steeper farther down toward the bottom. Forty miles above Lewiston the Clearwater receives the North Fork, which drains a large territory, its headwaters reaching the Montana-Idaho boundary line. This fork lies almost entirely outside the territory covered by this reconnaissance. Forty miles farther up the river again divides; the Middle Fork continues 25 miles due east in a deep canyon cut through the westerly ramparts of the Clearwater Mountains and then forks again. The northerly branch, Lochsa Fork, heads near Lolo Pass and flows through a canyon 4,000 feet deep, receiving few tributaries from either side. The southerly branch, called the Selway Fork, drains the larger part of the western slope of the Bitterroot Mountains and splits up in a maze of branching and winding streams. The South Fork, finally, which is considerably smaller, and has a less deeply trenched canyon, heads near the Salmon River divide, first flowing in a westerly direction, changing at Mount Idaho to a northerly course.

The grade of the rivers is by no means excessively great, that of Lochsa and Selway forks varying between 25 and 40 feet per mile through the main part of the mountains. Close to the Bitterroot Mountains the grade becomes much steeper and the branches finally head in the great glacial cirques on the west side of the Bitterroot Range.

A peculiar feature throughout this region is that the rivers receive their largest tributaries from the south and that near their headwaters they have a tendency to bend to the south.

The Salmon River throughout this region flows in a deep canyon without appreciable amounts of bottom lands. This trench is less split up into branches and is even more impressive than that of the Clearwater, its sides being also more precipitous. From Shoup to Freedom, a distance of at least 130 miles, the canyon is from 4,000 to 5,000 feet deep and about 10 miles wide, and has cut through the great uplifts of pre-Tertiary rocks. From Freedom to its mouth the river flows for about 60 miles through a high plateau of Columbia River lava, older rocks being, however, generally exposed in the bottom of its canyon.

Snake River is the main trunk stream of this region, and here forms the boundary of Idaho on one side and Oregon and Washington on the other side.

It first receives the Salmon and, 40 miles farther down, the Clearwater. Throughout this region it flows through the Columbia lava plateau in a deeply incised canyon, usually with steep sides. This canyon "may be divided into two well-defined sections of about equal length. Throughout the southern half of its course the river is in the bottom of a magnificent canyon with deeply sculptured walls 4,000 feet high and about 15 miles wide at the top. The rim rock at the crest of the walls on each side is formed by the edges of a lava sheet which immediately underlies the surface of Craig Mountain and of another similar plateau in Oregon. The northern half of the portion of the canyon referred to is narrow, being perhaps 2 miles wide at the top, and is cut in nearly horizontally bedded lava sheets; the walls, which are so steep that they appear vertical, are about 2,000 feet in height."[a] (Pl. XIII.)

GEOLOGY.

GRANITE.

From all available information granitic rocks of massive structure prevail over the whole extent of the Clearwater Mountains. Gneisses appear only at Pierce, Elk City, Salmon Mountain, and Horse Creek in areas of moderate extent, everywhere shattered and intruded by granite dikes. The Clearwater granite is, in general, a light-gray granular rock containing biotite and some muscovite, quartz, orthoclase, and a large quantity of oligoclase, indicating a relationship to the quartz-monzonites. The grain varies from coarse to medium, and larger orthoclase crystals up to 1 inch in diameter are sometimes present. The quantity of soda-lime feldspar is also variable, and in a few places the rock is a normal granite. Pegmatitic dikes are not very common except in some portions of the gneissoid areas. The structure is usually massive, though an incipient schistosity is noted in some places, being especially frequent in dikes contained in the gneiss.

From Lolo Pass, St. Marys Peak, and the summits at the head of Mill Creek the white outcrops of vast granite areas are seen far to the west. The Grave Peak complex and the whole of Lolo Ridge are probably of granitic character. The rock from Sherman Peak, on the western slopes, is a normal rock with much biotite and a little muscovite, orthoclase, and microperthite and a large quantity of oligoclase. A few smaller feldspar prisms are labradorite. Near Pierce, on Oro Fino Creek, the granite is slightly schistose and dioritic modifications occur near the contact of the true gneisses, beginning a short distance below the town.

The Crags form the last high projecting point on the Lochsa-Selway divide. This rock, like the one just described, is closely allied to the quartz-monzonites.

---

a Russell, I. C., Water-Sup. and Irr. Paper No. 53, U. S. Geol. Survey, 1903, p. 63.

Orthoclase and perthite are present in smaller quantity, but there is much oligoclase; biotite occurs in small foils.

The complex of Bilk Mountain was not visited, but from appearance, reports, and photographs by Mr. Lippincott granite is here the prevailing if not the exclusively occurring rock. (Pl. XI, B.)

At Lost Horse Pass, described above (p. 44), the granite shows evidence of schistose structure and is cut by many joint planes. Granitic rocks also prevail for a long distance west of that pass.

On the Salmon River divide granite begins on Horse Creek, and is cut by many dikes of granitic and dioritic porphyries. The rock is here of the normal Bitterroot type but varies in grain. West of the gneiss area of Salmon Mountain light-gray biotite granite prevails again, but contains many sharply defined inclusions of gneiss and occasional dikes of granite-porphyry. On the ridge east of Little Salmon River the granite has a slightly gneissoid structure. From this point to Ryans Meadows on the Red River, where the wagon road to Elk City begins, granite prevails. A specimen 4 miles west of Little Salmon Meadows is a coarse biotite-granite, orthoclase prevailing over the plagioclase, which seems to be an andesine. In the scant outcrops inclusions of micaceous and often contorted gneiss are common, indicating, perhaps, the vicinity of a larger gneiss area. Many of these gneiss fragments are injected in the small pegmatitic dikes parallel to the schistosity. After the large gneiss area of Elk City has been traversed the road leads down from Summit House to Harpster on the Clearwater. Granite adjoins the gneiss here and occupies the western slope toward the river down to an elevation of 3,800 feet, where the granite borders with intrusive contact against black slates and limestone of probably Triassic age (p. 73).

Granite occupies large areas again on the head of Crooked River. A specimen from the Homestake mine is a coarse biotite-muscovite rock, containing abundant microcline and a small amount of oligoclase. It is clearly a typical granite, comparatively rare in this region.

On the other hand, at Buffalo Hump the light-gray granular rock contains very little muscovite, very abundant oligoclase, and probably also andesine; it is evidently a quartz-monzonite. The Florence granite is similar, while at Warren (south of Salmon River) there is much muscovite and a smaller quantity of soda-lime feldspar.

GNEISS.

The gneiss areas lie embedded at various places in the granite, dikes of which are abundant in the schistose rock. Ordinarily it is a biotite-microcline-gneiss which evidently is derived from a granite; crushing and folding on a small scale are very frequent. The dip, usually moderate, is characteristically variable.

Beginning on the north a series of coarse micaceous and hornblendic banded gneisses appear on Oro Fino Creek just below the mining camp of Pierce (p. 102). Their strike is N. 70° W., the dip being nearly vertical. South of Pierce on Musselshell Creek and thence up to Sherman Peak granite prevails, but smaller areas of gneissoid rocks, probably inclusions, are encountered in places.

South of this begins the greatest gneiss area of this region, extending 40 miles long and 20 miles wide across the Middle and South forks of the Clearwater. Most common is an often crumpled biotite-microcline-gneiss with characteristically variable strike and dip, the latter from 20° to 90°. Coming from the east on the Nez Perce trail this gneiss is first met at Ryans Meadows and continues to Elk City, 15 miles distant. It is coarse, rich in mica, and often crumpled perpendicularly to the strike. The gneissoid structure is chiefly caused by the parallel arrangement of the mica foils, to less extent by the elongation of other constituents. Microscopically it consists of much microcline, orthoclase, and quartz, with less amount of plagioclase of medium acidity. There is further biotite with some hornblende, titanite, and epidote, the latter seemingly primary. Sometimes the mica foils curve around the white constituents, giving the rock an ocellar structure. It is well exposed in the bed rock of American Hill placer mine (p. 93) and here contains bands of fine-grained very micaceous rock and many narrow pegmatite dikes parallel to the schistosity. Dikes and gneiss are greatly contorted and crumpled. At Buffalo Hill placer mine the bed rock is similar gneiss but is cut in all directions by granite dikes. On Crooked River gneiss with micaceous streaks continues to a point 2 miles below the Badger mill, where it is replaced by intrusive reddish granite. Similar micaceous gneisses prevail between Elk City and Newsome and continue, cut by granite dikes, to a point 2 miles west of Mountain House, on the main road, where it is cut out by intrusive granite. Three and one-half miles west of the same place a narrow wedge of gneiss crosses the road, but granite again sets in and continues to near the foot of the grade to the Clearwater River.

The Middle Fork cuts across this gneiss area from Stuart eastward to beyond the Lochsa and Selway forks. Six miles above Stuart granitic gneiss begins; 8½ miles above are greenish schists striking N. 30° W. and dipping 70° WSW. Then follows micaceous gneiss up to the forks of the river. The point of the ridge between these is of gneiss, but 6 miles west of the Crags the trail crosses into the granite which forms the Crags. Near the contract the gneiss is cut by dikes of granite and pegmatite.

East of the Elk City gneiss area the granite contains many gneissoid inclusions. No large areas are, however, encountered until the great reef of Salmon Mountain is crossed. This consists chiefly of gneiss, extending probably all the way from Caseknife Mountain on the edge of the Salmon River Canyon north to McGruder Mountain. The whole mass is a banded and contorted micaceous gneiss, striking northeast and dipping steeply northwest. A specimen from the top of Caseknife Mountain is a gray and medium-grained gneiss consisting of allotriomorphic quartz, orthoclase and andesine with many foils of dark-brown biotite. Granitic injections of aplite, hornblendic granite, granite-porphyry, etc., are very extensive.

Good exposures were seen by Mr. Stose in the glaciated gulch leading up to Caseknife Mountain from Salmon River on the southeast. Numerous slipping planes suggest movement along the schistosity. In places the gneiss is much folded, quartz veins crossing the schistosity partaking in this movement. Intrusive granite dikes are common and have been made gneissoid parallel to the schistosity.

On the ridge eastward between Salmon Mountain and Horse Creek granite prevails but contains very sharply defined and broken shattered fragments of gneiss.

The most easterly area of gneiss begins on Horse Creek and extends up to the divide between Salmon and Bitterroot rivers, including the mining district of Mineral Hill (p. 90). This area which may possibly connect with that of Salmon Mountain along the deep canyon of Salmon River, borders against intrusive granite on the west, and probably underlies the sedimentary series of the head of the South Fork of the Bitterroot on the east. The gneiss is often coarse and dark gray and shows ocellar structure due to the curving bands of biotite. It consists of biotite, quartz, much microcline, and a small amount of plagioclase. Many bands are highly micaceous and sometimes contorted. The strike of this gneiss at the Horse Creek mines is north and south; the dip 40° E. The poor outcrops rarely permit good determinations, and a secondary cleavage often makes it difficult to find the true schistosity. Many dikes of granite-porphyry and granite cut this series.

SEDIMENTARY ROCKS.

Only few and small areas of sedimentary rocks have thus far been noted in the Clearwater Mountains, granite prevailing everywhere with smaller areas of gneiss. The probably pre-Cambrian areas of Lolo Fork and South Fork of Bitterroot River apparently do not enter the watershed of the Clearwater River, though continuing close up to its limits. On the western side a small area of dark-gray, fine-grained quartzites is found on Relief Creek between Crooked River and Red River, south of Elk City. Nothing is known as to their relations with the surrounding gneisses. Pebbles of this rock are very common at the Buffalo Hill hydraulic mine on American River (see p. 94). The only other sedimentary rocks encountered form irregular inclusions in the granite of Buffalo Hump (see p. 98). They are slates, quartzites, and limestones of unknown age, shattered by granitic dikes and greatly contact metamorphosed to mica-schists, hornfels, and garnetiferous products. Many of these rocks contain microscopic tourmaline.

TERTIARY GRAVELS.

The great flows of the Columbia River lava dammed the old drainage ways to present elevation of about 3,000 or 3,200 feet. The natural result was heavy accumulations of gravels in the lower reaches of the Clearwater streams. Remaining benches of such gravels are beautifully developed in the Elk City mining district (p. 91), where the highest of these attains an elevation of 4,500 feet. Similar conditions existed, closer to the valley, at Pierce, where the highest remaining gravel benches reach elevation of 3,300 to 3,500 feet (p. 102). Remains of gravels are also found on the ridge between Lolo and Musselshell creeks at an elevation of 3,700 feet.

GLACIATION.

The large confluent ice sheet which once covered the whole of the Bitterroot Range has been described in some detail on a previous page (p. 51).

Within the Clearwater Mountains no evidence of such general and widespread glaciation is found, due of course to the fact that the mountains rarely attain an elevation equal to that of the Bitterroots. The glaciation in the Clearwater Mountains is confined to a few isolated localities, and evidences of its existence appear wherever any considerable area of country exceeds 8,000 feet in elevation.

Beginning on the north the first center of glaciation is met at Rhodes Peak on the Lolo Ridge, from which doubtless small glaciers extended in several directions. Going west another local center from which small glaciers radiated appears at Bald Mountain and Castle Butte. Finally, a few miles farther west on the slopes of Rocky Ridge three or four smaller glaciers appear to have descended the streams leading up to that peak.

The ridge between Lochsa and Selway forks of the Clearwater is rougher and higher than the Lolo Ridge, and Grave Peak was evidently the center of an extensive glaciation, the ice streams descending in all directions from it. It is probable that this ice field was connected by way of Diablo Peak with the main ice-covered area of the Bitterroot Mountains. The glacial lakes near Grave Peak are shown on Pl. XII.

Still farther west the Crags near the end of the ridge rise to an elevation of over 7,900 feet. Small glaciers doubtless descended from their summits.

On the southern divide between Salmon River and Selway Fork the effects of glaciation are only prominent around two centers. The first one is Salmon Mountain reaching 8,881 feet. From the broad shoulders of this mountain wide glacial cirques descend in several directions. The glaciated area here probably covered 20 or 30 square miles. The divide from this point eastward to the head of the South Fork of the Bitterroot is very high and local glaciers may well have existed in some places.

From Salmon Mountain westward for 40 miles the summits of the ridges are broad and show no effect of ice sculpture, although the elevation often rises above 7,000 feet. Forty miles due west of Salmon Mountain a high complex of granitic hills rises to a height of 8,810 feet, the highest point being called Buffalo Hump. A névé field covering several square miles existed here and glaciers extended eastward for several miles down into Lake Creek, which drains into Salmon River, and westward at least an equal distance into Sheep Creek, also draining toward the same stream.

West from the line drawn from Buffalo Hump north toward Rocky Ridge on the Lolo trail the country slopes rapidly toward the plateaus of Columbia River lava, and no evidences whatever of a former glaciation are observed.

FIG. 5.—Salmon River Canyon seen from diorite point at Holdenreid ranch, near Keuterville; looking southeast, upstream, toward White Bird. Diorite continues across river; granite hills of Mount Idaho in left background; horizontal basalt flows elsewhere.

THE LAVA PLATEAU.[a]

TOPOGRAPHY.

From the western heights of the Clearwater Mountains a rapid descent leads down to a volcanic plateau occupying vast areas in Idaho, Oregon, and Washington. Geologically, this is formed by a great number of thin, horizontal basaltic sheets, and the formation is called the Columbia River lava. Between the North Fork and Harpster (on the South Fork), a distance of 50 miles, this plateau has a marginal elevation of about 3,000 feet, and a strip of it, up to 10 miles wide, lies on the east side of the Clearwater River, largely open and grassy on the south, more heavily timbered toward the north. West of the 1,000 to 1,500 foot trench of the Clearwater the direct continuation of this lava plateau is found in the broad, open Camas and

---

[a] As this region has been described in considerable detail by Prof. I. C. Russell in Water-Supply and Irrigation Papers, Nos. 53 and 54, U. S. Geol. Survey, especially in regard to the Columbia River lava and the water supply dependent on it, the reader is referred to that valuable report for detailed information on these subjects.

Kamiah prairies, varying in elevation from 3,000 to 3,500 feet; the two are separated by the abrupt Lawyers Canyon, named after a Nez Perce chief prominent in the war of 1877. West of Camas Prairie rise the hills of the Cottonwood Buttes, points of older rocks projecting through the lava sheets to an elevation of over 4,500 feet. To the southwest of Kamiah Prairie and separated from the Cottonwood Buttes by a high gap (elevation 4,000 feet) extends Craig Mountain, a broad dome-shaped swell of the lava attaining nearly 5,000 feet. The latter overlooks the deep canyon which the Snake has cut through the lava sheets. The Cottonwood Buttes overlook the similar deep gash cut by the Salmon, fig. 5, both canyons being about 3,000 feet deep. The lava plateau covers the point between the two great rivers,

Fig. 6.—View north of Rocky Ridge on the Lolo Ridge over the Coeur d'Alene Mountains, showing the continued plateau character north of the Clearwater Mountains.

but has here an elevation of 4,000 feet. It also covers the whole adjacent part of Washington and Oregon, the elevations of the plateau increasing southward from 3,000 feet opposite Waha to 4,000 feet or more on the Grande Ronde and the Imnaha, important tributaries received by the Snake from the west. South of Grangeville the Clearwater Mountains rise boldly. North of the Kamiah Prairie the lava plateau descends to the Lewiston Plateau, which has an elevation of about 1,500 feet.

Near Oro Fino, the Clearwater River bends sharply and flows for 40 miles almost due west to Lewiston, where it joins the Snake. The latter river itself makes a sudden turn of 90° at Lewiston, and for considerable distance into Washington follows the east-west direction of its tributary. This east-west direction of the rivers stands in some causal connection with the bold scarp of the Union-town plateau which all along follows their north side; the general elevation of the latter plateau is 2,500 to 3,000 feet and the steep slope carries it down 1,500 to 2,000 feet to river level.

GEOLOGY.

PRE-TERTIARY FORMATIONS.

An examination of the many deeply cut canyons within the province of the Columbia River lava soon reveals the fact that the underlying surface of older pre-Miocene rocks is extremely uneven and that the bottoms of the main streams before the time of the lava flows were at a lower level than they are at present. Thus the old rocks are irregularly exposed in many canyons and again at other places west of the main slope of the Clearwater Mountains where they are partly or wholly surrounded by the volcanic flows which rise above the general level of their surface.[a]

The series of rocks exposed below the lava is of a very complex character and consists of slates, sandstones, limestones, greenstone-schists, gneiss, diorite, and granite, the detached exposures making the study difficult. It is pretty clear, however, that two or three systems of bedded rocks are present. The first and oldest is chiefly exposed on the North Fork and at Oro Fino and consists of true gneisses and mica-schists, sometimes with intercalated limestone; it bears the marks of extensive regional metamorphism and is probably very old, possibly pre-Cambrian, and should be classed with the gneisses of Pierce and Elk City. The second and younger series is exposed on the Cottonwood Buttes and in the great canyon of the Snake; it is distinctly sedimentary, containing lime-stone, sandstone, and black slates, much disturbed but of imperfectly known structure, and associated with it are large masses of greenstones partly schistose and tuffaceous.

A third division consists of the also distinctly sedimentary rocks of Salmon River Canyon, Mount Idaho, and Harpster, which continue up to this point from the Seven Devils with persistent northeasterly strike. This division, which may be identical in age with the second series, is believed to be of Mesozoic-Triassic age. Granite and diorite, forming a spur of the great area of this kind of the Clearwater Mountains, has been intruded on a large scale into these three divisions of sediments. A bay of these rocks extends from Pierce, crossing the Clearwater near Kamiah and Greers Ferry, to eastern part of the Cotton-wood Buttes, probably reaching the canyon of the Salmon west of Grangeville.

---

[a] For these projecting points of more ancient rocks, islands as it were in the lava floods, Professor Russell has proposed the name of "Steptoes," from a butte of that name north of the area here considered.

Of all the main underlying rock formations this is doubtless the youngest. An approximately correct map of the exposures below the lava may be seen on Pl. I.

Beginning with Snake River, the first exposure of pre-Neocene rocks is noted in the basalt canyon of the Snake River, 26 miles below Lewiston. It is a coarse light-gray gneiss, apparently a sheared granite with large pressed aggregate of orthoclase and quartz, surrounded by flaky films of biotite and hornblende. The exposure is one-half mile long and rises 250 feet above the river. On the Clearwater no exposures of old rocks are seen until 25 miles above Lewiston and 3 miles below Contact, from which place the exposures continue for 8 miles until a point above Oro Fino. In Bedrock Creek and Big Canyon, quartzites and greenstone-schists are exposed, seemingly identical with the Cottonwood Butte series and having a strike of N. 25° W., the dip being 30° E.

At the mouth of the North Fork, Professor Russell[a] found diorite, gneiss, and micaceous garnetiferous schists striking a little west of north and dipping 70° to 80° E. One and one-half miles above the mouth of North Fork the rock is a fine-grained gneiss with bands of whitish quartzite. The strike is N. 70° W., the dip 65° NNE. A short distance above the town of Oro Fino the rocks are micaceous quartzite and mica-schists, containing a coarsely crystalline limestone in well-defined beds 120 feet wide, which may be followed for 1½ miles southward, where it is narrow, and one-fourth mile northward. The strike is N. 25° E., and the dip 78° ESE. Two and one-half miles above Oro Fino the micaceous quartzites and mica-schists strike N. 55° W., dipping steeply southwest. Near the mouth of Ford Creek basalt covers the bottom of the canyon. South of Ford Creek, about Greers Ferry and the mouth of Lolo Creek, the prevailing rock is a massive granite or diorite, some of it at Greers Ferry being a very beautiful coarse pegmatite, consisting of plagioclase and hornblende. A few miles above Kamiah the granite disappears below the basalt, which now for 10 miles or until a few miles below Harpster, forms the bed of the river.

Small bed-rock exposures occur on the top of the plateau near Fletcher, on the head of Big Canyon, and on Mission Creek, where limestone has been observed by Professor Russell.[b]

The Cottonwood Buttes are, as stated above, a series of pine-clad hills projecting somewhat over 1,000 feet above the lava plateau of Camas Prairie. The western and highest part consists of prevailing greenstone-schists, alternating with massive greenstones and smaller areas of partly tuffaceous clay slate and quartzitic slates, the latter fairly being well exposed on the road from Cotton-

---

a Water-Sup. and Irr. Paper No. 53, U. S. Geol. Survey, 1903, p. 24.

b Water-Sup. and Irr. Paper No. 54, U. S. Geol. Survey, 1903, p. 120. Analyses are here given of limestones from Snake River, Mission Creek, and Oro Fino. All are practically pure carbonate of calcium, the last named only containing 3 per cent magnesia.

wood to Keuterville. The greenstones appear to be old lavas of andesitic or basaltic character. The strike is northeasterly, the dip being at a high angle and rarely determinable with exactness. Eastward the older rocks continue for several miles out into the lower country toward Greer Creek and Columbus, but is here principally granite and granite-porphyry. Immediately south of these buttes the plateau breaks off into the great canyon of the Salmon. A narrow and shallow rim of basalt flows, separates the old rocks of Cottonwood Buttes from this great trench, and from prominent points on the edge of the canyon a magnificent view is obtained (fig. 5). The rocks are here chiefly greenstone-schists with some massive dark and medium-grained diorite. The Cottonwood Buttes are recognized as a great reef of the sublava series continuing across the canyon with a southwesterly direction. Across the Salmon River the highest point of the old spur is only 2,000 feet above the sea and is covered with 2,000 feet of basalt.

In the Salmon Canyon bed rock is exposed for at least 30 miles between a point southwest of Keuterville and White Bird Creek. Little is known as to its character. Professor Russell states that below Keuterville there is much of a light-colored igneous rock. Limestone is reported from a point in the bottom of the canyon between Rocky Canyon and White Bird Creek.

At Denver, one of the lowest points on the Camas Prairie (elevation 3,125 feet) and situated on the divide between the Salmon and the Clearwater, bore holes have penetrated 240 feet of lava and struck, according to reports, a granitic rock at that depth, water under considerable artesian pressure being also obtained, rising to a point 150 feet below the surface and indicating a bed-rock depression with supply from the direction of the Cottonwood Buttes or Mount Idaho.

The bed-rock exposures along the Snake River above Lewiston have been described by Professor Russell,[a] and an abstract of his statement is given below.

The first outcrops of older rocks in the canyon occur 12 miles above Lewiston, west of Waha. A bluff of diorite-porphyry rises here on the east side of the canyon below the lava to a height of 3,300 feet above the river; a lower bluff of the same kind appears on the west side of the canyon. A few miles farther up, at the mouth of Grande Ronde River, the underlying rocks appear again with rough topographic form, the summits of the old topography attaining a height of 2,500 feet above the river and being covered by 1,500 feet of horizontal basalt flows. The rocks are thin-bedded limestones (with indefinite fossils), schists, and diorite. The dip is generally northward at angles of 40°–45°, sometimes also vertical. The rocks beneath the basalt are exposed in the lower 2,000 feet of the canyon all the way up to a few miles above where the Salmon River

a Water-Sup. and Irr. Paper No. 58, U. S. Geol. Survey, 1903, p. 26.

enters. They were examined by Mr. Russell at Cottonwood Creek a few miles below the mouth of the Salmon; he states that the stratified rocks are much disturbed but generally strike northeast-southwest with variable dip. Sandstone conglomerates, shale, and limestone were observed. Intrusive rocks, diorite, and diabase-porphyry also occur. Quartzite and diorite were noted near the junction of the Salmon and the Snake.

It remains to describe the interesting series of rocks from the canyon of the Salmon above White Bird, from Mount Idaho and from the South Fork of the Clearwater.

One hundred and ten miles south-southwest of Grangeville, at Huntington, Oreg., at the beginning of the great canyon of the Snake, first appears a series of slates and limestones with great masses of interbedded old lavas. These rocks, which continue for 30 miles along the canyon, have a northeasterly strike and prevailing northwesterly moderate dip. Their age is not positively known. Farther north this series is covered by Neocene basalts. The same, or at any rate, a very similar series appears at the Heald mining district and in the Seven Devils, Idaho, old surface lavas with intercalated slates and limestones prevailing. Fossils found in the Seven Devils indicate with certainty a Triassic age.[a]

Thirty miles northeast of the Seven Devils a similar series of sediments and associated igneous rocks is excellently exposed along the Salmon River Canyon,[b] near Freedom, the interval between these two locations being occupied by high, rugged ridges as yet unexplored. The prevailing series in this part of the Salmon River Canyon consists of clay slates and persistent beds of crystalline limestone with associated partly sheared greenstones (chiefly old surface lavas), having a general northeasterly strike and a southeasterly dip of about 45°. No fossils were found here. Granites and diorites, exposed in places below the horizontal lava sheets, and belonging to the great intrusive area of central Idaho, adjoin the whole series on the Idaho side from near Weiser, Idaho, the contact clearly having a northeasterly trend. The same contact is exposed east of Salmon River, as shown on Pl. IX of the Twentieth Annual Report of the United States Geological Survey, pt. 3, and also on Pl. I, this report. Extensive contact metamorphism marks the strip of slates adjoining the granite. In places, as above the great bend of the Salmon, 10 miles below where the wagon road to Florence crosses the canyon, gneissoid rocks are present along the contact in such volume as to suggest that they may possibly be part of an independent formation instead of contact metamorphic sediments.

aLindgren, W., The gold belt of the Blue Mountains of Oregon: Twenty-second Ann. Rept. U. S. Geol. Survey, pt. 8, Pl. LXIV, p. 681. The gold and silver veins of Silver City, De Lamar, and other mining districts in Idaho: Twentieth Ann. Rept. U. S. Geol. Survey, pt. 3, Pl. IX, pp. 87-89.

bLindgren, Twentieth Ann. Rept. U. S. Geol. Survey, pt. 3, 1900, p. 581.

Along the road from Mount Idaho to Florence, a few miles south of the former town, granite and some schists are exposed below the lava contact.

East of Mount Idaho, in the canyon of the South Fork of the Clearwater occurs a series of sediments which seem to be the direct continuation of those of the Salmon River Canyon. Mount Idaho is situated on the basaltic plateau, but in the canyon immediately east of it and 1,500 feet lower greenstone-schists, with bands of clay slate having a northeasterly strike, prevail. They are well exposed between the Dewey mine (p. 105) and Hinckley's ranch. A few injected granite dikes were observed in the schists. On the steeply rising hills east of the river at Hinckley's greenstone-schists first prevail, but 1,600 feet above the stream a strong and persistent belt of crystalline limestone is crossed, adjoined on the east by heavy masses of black and gray banded quartzitic slates.

At Harpster, 10 miles farther down the South Fork, dark-green, medium-grained diorite is exposed in the bottom of the canyon below the lava, and continues for a few miles beyond that place until the lava dips below the river level. Prospectors affirmed that the diorite continues for 4 miles above the Harpster bridge. Above this there is said to be an alternation of limestone, slate, and greenstone.

The most northerly exposure of this important northeasterly striking series is found on the road 5 miles east of Harpster. At an elevation of 3,850 feet, at the foot of a long grade descending over granite from Summit House and near the beginning of the lava plateau, appear outcrops of finely laminated clay slates striking N. 60° E. and containing well-defined dikes of granite. Two miles south of this point Mr. Johnson found a bed of pure white crystalline limestone 60 feet thick associated with decomposed 'greenish slates. This limestone would seem to be in the northeasterly continuation of the bed noted east of Hinckley's ranch; it had been utilized for quicklime, as indicated by a kiln built at the outcrop.

Northeast of this point the sedimentary series is replaced by an apparently far older gneissoid series. The similarity of this whole northeasterly striking series from Huntington to the South Fork of the Clearwater and the definite identification at the Seven Devils warrant the belief that it is throughout, largely at least, of Triassic age.

BASALT FLOWS.

Topographically and geologically basalt is the most important component rock of the high plateau spread out at the foot of the Clearwater Mountains. As already indicated, the plateau is chiefly built up of a great number of horizontal basaltic flows, to which Professor Russell has given the formation name of the Columbia River lava. On the rolling surfaces of the high prairies little is seen of this formation except the deep, fertile, dark-brown soil, to which it easily disintegrates.

Along the walls of the many abrupt canyons which trench the formation its composition is beautifully and clearly revealed. The cliffs and slopes, of somber dark-brown color, present a great number of horizontal, long, and persistent lines delimiting the individual flows and due to their varying resisting power toward erosion. Each flow is from 25 to 150 feet thick, the upper and sometimes also the lower part of each being more scoriaceous than the central part. Throughout the prevailing rock is a basalt,[a] though varying much, for while some flows are very glassy others are almost granular and have a diabasic structure. These differences are no doubt due to differing conditions of cooling. A columnar structure, due to vertical contraction joints, is observed in places.

In fresh specimens the basalt is nearly black, sometimes vesicular, more frequently massive. Typical specimens from the Clearwater Canyon, 4 miles below Contact and from the mouth of Potlatch Creek, are massive, black rocks of fine grain, and containing prisms of labradorite up to one-half inch long. In thin section they prove to be olivine-basalt of diabasic granular structure, but also containing between the grains some dark-brown glass. Porphyritic labradorite crystals are on the whole very common in many flows. The basalts from the plateau of the Columbia River lava near the Seven Devils, Eagle Creek Range, and Boise are entirely similar.

The only analysis of the Columbia River lava thus far available is a partial one, by Prof. Geo. P. Merrill, of a basalt from Walla Walla, and runs as follows:[b]

Analysis of basalt from Walla Walla, Wash.

SiO_2	47.35
Al_2O_3 Fe_2O_3	34.38
MgO	4.43
CaO	8.27
Na_2O	2.55
K_2O	1.33
Loss on ignition	.95
	99.26

The persistence in petrographic characters over large areas is noteworthy.

The age of the basalt flows would seem to be Miocene from the paleo-botanical evidence thus far available. (See p. 24.) Possibly Eocene eruptions make up a part of the formations hitherto grouped under the name of Columbia River lava.

---

[a] Russell, I. C., and Diller, J. S., Water-Sup. and Irr. Paper No. 4, 1897, p. 43. Lindgren, W., Twentieth Ann. Rept. U. S. Geol. Survey, pt. 3, 1900, p. 90–93; Twenty-second Ann. Rept. U. S. Geol. Survey, pt. 3, 1902, pp. 592–593; Geologic Atlas U. S., folio No. 45, Boise, Idaho.

[b] Water-Sup. and Irr. Paper No. 53, U. S. Geol. Survey, 1903, p. 44.

Intercalated between the basalt flows are occasionally found thin beds of white volcanic dust of rhyolitic character. Professor Russell mentions such occurrences from each side of Potlatch Canyon; beneath the Candle Rocks, on the left bank of Snake River below Lewiston; near Swallow Rock, between Clarkston and Asotin; also on Asotin and Captain John Creek in the canyon of the Snake above Lewiston.[a]

Waterlaid beds, consisting of clays and sands swept down from the surrounding mountains, occur in some places between the lava flows. Basaltic tuffs are practically absent. Professor Russell states that on Small's ranch, on the Lewiston Plateau, the drill encountered 60 feet of sand and 90 to 100 feet of clay below 210 feet of basalt, these sediments being again underlain by basalt. These sedimentary, intercalated beds can be traced over a large extent of country in the northern and western portions of Nez Perce County and adjacent portions of Washington and Oregon. They probably extend to the foothills of the Bitterroot Mountains, but as there are other similar layers in that region they have not been identified east of Craig Mountain.[b] At Cottonwood, on the Camas Prairie, gravel 56 feet deep has been penetrated below a surface sheet of 60 to 70 feet of basalt.

Along the canyon of the Clearwater River, from Stuart to Peck, there are indications of a thick sheet of sandstone and allied rocks at an elevation of about 800 feet below the general surface of the plateau. The true relations are often obscured by landslides. The same, or a similar complex, 200–300 feet thick, is shown throughout the length of Oro Fino Creek from its mouth to near Pierce along the walls of the canyon. These beds begin about 750 feet below the surface of the plateau, as along the Clearwater Canyon landslides are very common. Again, sandstone and carbonaceous slate occur at the same level on Little Canyon Creek, draining a part of the broad plateau south of Oro Fino. Finally, on Potlatch Creek, which joins the Clearwater from the north about 18 miles east of Lewiston, a moderate thickness of sedimentary beds and volcanic dust appears in the basaltic series, 3 miles above the mouth of the canyon.[c] "The Uniontown Plateau is underlain all along its southern border, and apparently throughout practically its entire area, by horizontal sheets of basalt, which include at least one important bed of clay, sand, and gravel, at a depth of about 160 feet below the surface."[d]

STRUCTURE AND PHYSIOGRAPHY OF THE PLATEAU.

Orogenic disturbances.—Like any other series of horizontally deposited beds the Columbia River lava is likely to be deformed into folds, and like any other

---

a For analysis of this see Water-Sup. and Irr. Paper No. 53, U. S. Geol. Survey, 1903, p. 34.
b Water-Sup. and Irr. Paper No. 53, U. S. Geol. Survey, 1903, p. 35.
c Ibid., pp. 38–39.
d Ibid., p. 57.

terrane it may of course be broken and faulted by orogenic forces. Comparatively little such folding or faulting has taken place, indicating a general absence of tangential compressive stresses. This seems to hold good, not only of the lava plateau, but also of the higher and older largely granitic plateau of the Coeur d'Alene, Clearwater, and Salmon River Mountains. The whole of central Idaho should perhaps be regarded as a rigid crust block, but little folded or faulted since Mesozoic time.

Some disturbances of the Columbia River lava have, however, occurred in this region. Along the foot of the Clearwater Mountains the top of the lava flows is generally found at an elevation of 3,000 feet above the sea. Practically the same elevations obtain on the Camas and Kamiah prairies and on the Uniontown Plateau, and this level may be regarded as the surface of the sea of molten rock. Fifteen hundred feet lower than the general level is the Lewiston Plateau, bordered on the north by the Clearwater escarpment. Up to 1,500 feet higher than this level, is Craig Mountain, a long, broad lava swell, extending in a west-northwest direction between the Snake and the Clearwater. (Pl. I.) Both of these features Professor Russell regards as monoclinal folds.[a] The Lewiston Plateau is part of a synclinal trough, bordered on the north by the monoclinal fold of the Clearwater escarpment (Pl. XIV). Craig Mountain is a gentle anticline separating the Lewiston Plateau from Camas Prairie. (Pl. I.)

Epeirogenic disturbances.—In a previous paragraph it has been shown that the combined ridge lines of the Clearwater Mountains (and also of part at least of the Salmon River Mountains south of the stream of the same name) would form a broad plateau surface, with an average elevation of about 7,000 feet. In places this plateau was depressed to 6,000 feet, while in other places irregular groups of hills rose to 1,000 feet above the average level. A slight westward slope probably existed and was the determining factor in the general plan of the rivers. This surface was evidently a surface of erosion, an old topography reduced from still older mountain systems, and lifted, like the rest of the great Idaho mountain mass, by epeirogenic processes long before the Miocene epoch to the position of a high plateau in which rejuvenated rivers could begin new and active processes of canyon cutting. The rather sudden descent from this old plateau to the later one of the Miocene lava is noteworthy. It can scarcely be due to erosion, but is rather due to deformation of the crust as suggested in the case of the similar steep descent of the plateau of the Boise Mountains to the lower valley of the Snake River.[b] The general character of this plateau between Florence and Buffalo Hump, and its contrast against the deep canyon of the Salmon River, is clearly shown on Pl. IX of the Twentieth Annual Report of the United States Geological Survey, pt. 3.

---

a Water-Sup. and Irr. Paper No. 53, U. S. Geol. Survey, 1903, pp. 54–61. b Geologic Atlas U. S., folio No. 45, Boise, Idaho.

By following the contact line of the Columbia River lava and the older formations of the Clearwater Mountains, it soon becomes evident that the basalt flooded the foothills to an approximate elevation of 3,000 feet; further, that the foothills of the Clearwater Mountains which it covered often had a sharply accentuated topography, and that the lowest points of the main streams of the old pre-Miocene system were considerably lower than the bottom of the drainage of to-day. The present streams of the region, the Clearwater and the Salmon, flow through the mountains in canyons which have a depth of from 3,000 to 5,000 feet. The course of the upper and main parts of these streams could have been but little influenced by the floods of basalt even had they filled their canyons up to a level of 3,000 above the sea. They would in short time have reexcavated their canyons in the lava covering the bottom, and the main features of the great canyons would not have been changed. Only the lowest river courses, below a point where the lava flooded the divides of the canyons, would have been different.

We thus come to regard the whole canyon system of the main fork of the Salmon above Freedom and the similar system of the Clearwater above Stuart and Harpster as a very old one, laid out as a constructive drainage on the slightly westward sloping plateau of the Clearwater Mountains. Its canyons were excavated almost to their present depth in Miocene time, before the Columbia River lava had flooded their lower reaches. Probably the grade was somewhat steeper, the post-Miocene erosion having succeeded in gradually reducing their gradient. The present steep slopes of the canyon walls have been retained, and are most emphasized near the stream lines because of the damming of the lower reaches to elevations of 3,000 feet during the Miocene.

Below the point where the lavas covered one or both of the confining ridges of the canyon the new river course would probably leave the direction of the old valley and follow the chance inequalities of the lava plain. A second series of constructive drainage would thus be outlined, and of this kind is the canyon of the Clearwater below Harpster, the canyon of the Salmon below Freedom, and the great canyon of the Snake River in practically its entire course between Huntington, Oreg., and its junction with the Columbia. In the Clearwater River, lava is first encountered in the bottom of the stream a few miles below Harpster. Consulting Pl. I, we may conclude that the Miocene canyon turned westward from this point and continued below Lawyers Canyon. From here it either extended north-westerly below Big Canyon toward Lewiston, or, less probably, followed the depression between Craig Mountain and Cottonwood Butte in a southwesterly direction. Beyond Big Canyon all is covered by basalt, and the direction of the old river is consequently impossible to determine from surface indications. That an important tributary joined this river below the Lewiston Plateau is indicated by the deep depression, bringing lava to the bottom of the Clearwater Canyon between Greers Ferry and Oro Fino.

In a similar way the pre-Miocene trench of the Salmon River must have continued below the lava plateau between the Snake and the Salmon, and probably crossed the present canyon of the Salmon a few miles above its mouth and the Snake River Canyon between Grande Ronde and Waha.

As stated above, the canyon of the Snake River is of Miocene date and differs entirely from the pre-Miocene valley through which the drainage of the great Snake River Valley found its outlet. The relations of lava and bed rock along the course of the Snake River Canyon from Huntington down indicate that this pre-Miocene trench began near Weiser, Idaho, crossed the present canyon about the mouth of Pine Creek, and continued northwesterly between the Seven Devils and the Eagle Creek Range.

Wherever along canyons the contact lines between lavas and older rocks reveal the old topography its accentuated character is prominent. It is thus at Waha, at the mouth of the Grande Ronde, at Freedom, and at Harpster. In the lower reaches of the rivers the old channels were cut deeper down into the bed rock than the present trenches have been enabled to do, although they have been excavated so deep that at present the river at Lewiston is only about 725 feet above the sea. This leads us to the conclusion that the region of the Columbia River lava along the mountains of Idaho has been depressed, and that it is a field of subsidence of epeirogenic character. This hypothesis I first formulated in a report dealing with the general geology of south-central Idaho[a] and repeated it in a later report on the Blue Mountains of Oregon.[b] Professor Russell has also arrived at the same conclusion.[c]

The problem would be less difficult if the surface of the lava everywhere approximately followed the 3,000-foot contour, as it mostly does along the foot of the Clearwater Mountains, but such is not always the case. On Mount Idaho and on the plateau between the Salmon and the Snake the basalt flows rise to elevations of 4,000 and 5,000 feet. Near the Seven Devils and the Eagle Creek Range they even attain 6,000 and 7,000 feet. Professor Russell[c] reasons that because the lava sheet is in places 4,000 feet thick the top sheet must have had at least that elevation above the sea, unless the old land surface was below the level of the sea, which is admittedly improbable. From this it might be expected that, since the lava sheets in eastern Washington and Oregon and adjacent portions of Idaho are still horizontal, the surface of the plateaus thus formed should have an elevation of at least 4,000 feet above the sea. Instead, we find in numerous instances that the general elevation of the plateaus underlain by essentially horizontal lava

[a] Twentieth Ann. Rept. U. S. Geol. Survey, pt. 3, 1900, p. 96.
[b] Twenty-second Ann. Rept. U. S. Geol. Survey, pt. 3, 1902, p. 598.
[c] Water-Sup. and Irr. Paper No. 53, U. S. Geol. Survey, 1901, p. 54.

sheets is between 3,000 and 3,300 feet. This difference between the present elevation of the locally undisturbed plateaus and what is assumed to have been their original position indicates a subsidence of about 1,000 feet, which may have gone on during the time in which the lava was being extruded or may have occurred in part or wholly since the last sheet was spread out. Proceeding on the same lines, Professor Russell concludes that where slightly inclined strata of lava are found which stand high above the general level of 4,000 feet they have been upraised, and vice versa. The above line of reasoning starts with the assumption that the basalt spread out under hydrostatic conditions. This can not be unqualifiedly admitted to be true; for it seems well possible that near the points of eruption the sheets may have been piled up to gently sloping accumulations considerably thicker than elsewhere.

It is certain, however, that the course of the whole canyon of the Snake River and of the lower canyon of the Salmon River is difficult to explain except on the assumption that they were originally laid out over an approximately level plateau which has been gradually raised simultaneously with the cutting of the canyon, thus giving them the character of antecedent streams. The course of the Snake River Canyon has been established across the high plateaus at present 5,000 to 7,000 feet above the sea, whereas if no uplifts had occurred other directions would seem to have offered a far more natural outlet for the stream.

In conclusion it is believed that the region of the Columbia River lava has suffered a general subsidence accompanied by more localized uplifts.

PLACE OF ERUPTION OF THE BASALT.

There is no evidence of the existence of ash cones or tuff volcanoes within this part of the Columbia River lava formation, and it is thus more probable that the eruption of the fluid rock took place, without explosive action, from fissures from which the basalt flowed readily and continuously. No dikes of basalt were, however, observed in the foothills of the Clearwater Mountains, and we are thus forced to seek the focus of the eruption elsewhere. Only one local center of eruption has thus far been discovered. This locality is in the Eagle Creek Range, Oregon, and has been described in detail in a previous report.[a] A perfect network of basalt dikes here intersect the schists and granites near Cornucopia at elevations of about 7,000 to 8,000 feet, and immediately below this place the lava plateau begins.

While the total thickness of the sheets sometimes reaches 4,000 feet, the average depth of the Columbia River lava is ordinarily much less in the region here described, 3,000 feet being probably the maximum and 2,000 feet possibly the average.

---

[a] Twenty-second Ann. Rept. U. S. Geol. Survey, pt. 3, 1902, pp. 740–745.

COEUR D'ALENE MOUNTAINS.

The geology of these mountains is practically unknown, and this may be the excuse for publishing the following brief notes taken during 1899, on a trip from Lewiston to Spokane and thence across to Montana by way of Mullan Pass. A short stay was made at the Coeur d'Alene mines, and such notes regarding these important lead deposits as are available will be found on page 108.

Going north from Lewiston on a branch of the Northern Pacific Railway one soon ascends the Uniontown Plateau, which has an elevation of 2,500 feet, and the route continues for 50 miles northward just west of the Washington-Idaho line, over a rolling open country of Columbia River lava, well cultivated and forming a part of the rich wheat lands of the "Palouse country." Spurs of older rocks project from the Coeur d'Alene Mountains and points of the same in places reach up above the general surface of the lava plateau. Of the former kind are the Cedar Mountains, near Moscow, Idaho, consisting of quartzite and granite, and the Tekoa Spur from the Moose Creek Mountains, which consists of slate. Of the latter kind are several quartzite points north of Pullman, as well as the Steptoe and Kamiack buttes described by Russell.[a] Fifteen miles from Spokane the road crosses an area of very decomposed granite, locally used for the manufacture of fire bricks, and descends to the Spokane River, which near Spokane is flanked by basalt flows. These are underlain by fluviatile or lacustrine clays, which are probably younger than the Columbia River lava.

From Tekoa a branch road runs up to Coeur d'Alene Lake and thence eastward across Mullan Pass into Montana. Along this line the formations seen are as follows:

After crossing slates and quartzites, a spur from the small complex known as the Moose Creek Mountain, the basalt begins again near Coeur d'Alene Lake and rises to bluffs forming a plateau 500 feet above the water (elevation about 2,500 feet). The Columbia River lava continues up the Coeur d'Alene River a few miles east of the lake, and then begins a sedimentary series of great thickness, which continues with west-northwest or northwest strike almost uninterruptedly to Missoula, Mont. The mountains consist of a maze of steep, narrow ridges, 1,000 to 3,000 feet high, the level crest lines of which have an elevation of from 5,000 to 6,500 feet. This region should be regarded as a very extensively dissected plateau. At first banded quartzitic slates prevail. North of Wardner the prevailing strike of the slates is N. 62° W. and the dip about 50° NNE. At Wardner the rock is a heavy-bedded gray quartzite, of medium grain; the bedding planes strike N. 70° W. with a dip of 60° SSW. Quartzite continues up to Burke and

---

[a] Water-Sup. and Irr. Paper No. 34, U. S. Geol. Survey, 1900.

Mullan, occasionally with narrow intercalated bands of clay slate. A mass of granite breaks through the quartzite between Ninemile Creek and Canyon Creek at Burke and is believed to stand in some genetic connection with the ore deposits. Possibly this granite continues northward to Murray, where gold placers and veins have been found. From Mullan Pass the railroad follows down the St. Regis de Borgia River for 30 miles, where it joins the Missoula; from this point it skirts along the Missoula River for 60 miles to a point near the town of the same name. The general direction of the road is east-southeast from Mullan Pass. Along this whole distance the prevailing rock are banded quartzites and slates, which strike northwest and dip variably but generally at 30° to 50° NE. Interbedded with the quartzites 30 miles west of Missoula are purple slates similar to those found on Lolo Fork.

The whole series from Missoula to Coeur d'Alene Lake is evidently folded and compressed, and its thickness must be very great. Its age is unknown, but may well be pre-Cambrian.

ECONOMIC GEOLOGY.

GENERAL FEATURES.

Character of mineral deposits.—The valuable mineral deposits occurring in the area described in this report chiefly consist of fissure veins containing gold, together with associated placers, derived from the disintegration of the veins. Deposits containing lead and copper, usually also silver, occur also in several scattered places. Coal of an inferior quality has also been found in the upper Bitterroot Valley and in the lower Clearwater drainage. The lead-silver veins of the Coeur d'Alene Mountains are outside of the limits of this reconnaissance.

Distribution of deposits.—The metalliferous deposits are grouped in two belts. The first is along the western side of the Bitterroot Mountains, chiefly in Montana; the second is along the western foot of the Clearwater Mountains, in Idaho. The deposits of each of these two belts are again principally grouped in two regions forming the four corners of the mountain area involved, while the central part of the Clearwater Mountains appears to be practically barren. The four metalliferous areas are distributed as follows: The first occupies the lower Lolo Fork and the northern end of the Bitterroot Mountains; the second is found on the headwaters of the South Fork of Bitterroot River, and reaches over into Idaho, connecting with the mineral belts at Shoup and Gibbonsville; the third, the most important area, includes Elk City, Buffalo Hump, Dixie, and Florence, as well as numerous places along the South Fork of Clearwater River; the fourth area centers in Pierce, but also extends to the headwaters of Lolo Creek on the south and to the North Fork of the Clearwater on the north.

Character of ore.—The primary deposits are almost exclusively fissure veins and with them are associated extensive placers of an age ranging from Miocene or Pliocene to recent. In the northern Bitterroot Mountains and on Lolo Fork veins occurring in pre-Cambrian (?) schists chiefly contain copper, lead, and silver, although some gold is also found on Lolo Fork. The Curlew mine at the eastern foot of the Bitterroot Mountains contains argentiferous galena, and is located on a fissure between limestone (pre-Cambrian?) as the foot wall and, according to accounts, Pleistocene valley gravels as a hanging wall. The mine is not worked at the present time. On the upper Bitterroot River veins cutting porphyry likewise chiefly carry copper and silver, while argentiferous galena is also known from the Monitor mine, worked on a vein in gneiss on the divide between Bitterroot and Salmon rivers. Gold-bearing gravels have been mined for many years on Hughes Creek. Southward this belt connects with the gold-bearing deposits at Gibbonsville and Shoup. The rocks at Hughes Creek and Gibbonsville are pre-Cambrian (?) quartzites and slates.

West of these districts extends a wide granite area which as far as known is barren of mineral deposits. There can be no doubt that the Clearwater drainage was very thoroughly prospected for placers during the early days of mining, but outside of the South Fork very little of value has been found. In the upper part of the mountains the glaciation would naturally have swept away any placer deposits which may have existed, and in this denuded portion it is not impossible that veins may be found. Nothing of much value has been encountered up to the present time. A large vein containing silver is reported to occur on Rhodes Peak north of the Lolo trail. Along the Salmon River the conditions are probably more favorable and prospecting in the isolated region between Dixie and Shoup might possibly develop something of value.

As stated before, the western belt chiefly contains gold; only a few scattered copper deposits are known. The placers of Elk City and Florence are well known in the history of Idaho and are still worked to some extent. Veins which furnish the material for these placers are known to occur in all these localities; the principal mining districts are Florence, Dixie, Elk City, and Newsome Creek. The veins, chiefly occurring in gneiss, are almost exclusively of quartzose character and contain from 1 to 10 per cent of sulphurets, besides more or less native gold. The Buffalo Hump district, discovered in 1898, is situated on the high divide between the Clearwater and the Salmon, and contains in granite and slate many strong quartz veins, with a varying percentage of free gold and auriferous sulphides. Active work is in progress there at the present time. The northwestern mineral-bearing area contains placers along Lolo Creek, Musselshell Creek, and Oro Fino Creek. Many quartz veins of similar character to those from the southwestern belt are also worked in these districts. They are generally incased in schists; more rarely in granite. Veins of sulphide ores, containing gold and copper, occur in amphibolite close to Mount Idaho.

History and production.—The deposits on the eastern slope of mountains have not proved of great importance and have chiefly been discovered and worked at a comparatively recent time. The total production of all the mines on this side of the mountains probably does not exceed $1,000,000, of which the larger part has been derived from the Curlew mine on the north and from the placers of Hughes Creek, near the head of Bitterroot River. The important gold belt on the western slope was discovered about 1860, and very actively worked during the following years. Oro Fino, or Pierce, is reported to be the earliest discovery in Idaho. It was found in 1860, and during that season twenty-five men wintered there. The gravel near Pierce was not remarkably rich, but paid fairly well in 1861 and 1862.[a] In 1874 Pierce produced $70,000. Soon after this, however, the discoveries in Montana drew most of the miners away from this place, and in 1867 but little mining was going on. Since that time, however, the placers and quartz mines have been worked each year, although in a somewhat desultory manner. The total production is impossible to ascertain, but probably has not exceeded a few million dollars.

During late years placer mining has been carried on both in the low stream gravels and on the benches. There has also been a considerable activity in quartz mining, and several small mills have been built. The output of the placer mines in 1902 is estimated at $30,000, and that of the quartz veins at the same amount.

Elk City and vicinity proved to be of greater richness. Few quartz mines have been worked there, practically the whole production being derived from the placers. In 1863 or 1864 the white miners began to leave this field, which they considered about worked out, and for nearly thirty years there were only two or three of them left in the district, which was almost entirely turned over to the Chinese. In 1892 the white miners began to come back, and the Chinese simultaneously disappeared, very few of the latter being left now. A certain amount of placer work is still done in this vicinity each year, chiefly on bench gravels. The bars of the Clearwater River, which were worked extensively during the early days, are still occasionally washed. Regarding the total output of Elk City no satisfactory figures are available, but not unlikely the production amounts to between $5,000,000 and $10,000,000.

After the first few years of abundant production the output fell rapidly. In 1874 Elk City (including Newsome Creek and Clearwater station) produced $120,000. From 1882 to 1887 the Elk City district produced from $35,000 to $73,000 per annum. During the recent years the output has again increased, due

---

[a] Browne, J. Ross, Report on Mineral Resources, Washington, 1868.

to the introduction of dredging and hydraulic operation, and during the last years has probably been from $20,000 to $40,000. Very similar were the conditions during later years in Florence, which camp has been described in a previous report.[a] The total output of Florence was, however, considerably larger than at Elk City.

Florence, Warren, and Elk City are situated in Idaho County. According to the mint reports this county has, since 1880, produced an average of $200,000 per annum, or a total of about $6,000,000. Something like one-half of this amount probably comes from Warren, leaving $3,000,000 for the two last camps for the last twenty years. Pierce is located in Shoshone County.[b]

It is somewhat surprising that in spite of the recent activity in prospecting and working quartz veins the production of Idaho County should have decreased during the last few years. The mint reports give for Idaho County:

Production of precious metals in Idaho County, Idaho.

1895	$243,700
1896	155,350
1897	236,500
1898	203,500
1899	166,000
1900	152,000
1901	161,500
1902	157,023

Geological relations.—Nearly all of the vein deposits occur in granite-gneiss or metamorphic slates and schists; the prevailing strike of the veins seems to be in an east-west direction. The granite, which is the prevailing rock, represents the northward continuation of the great area of central Idaho north of Snake River. Gold-bearing veins occur both within this area and along its contacts with the surrounding older sedimentary rocks.

Within the region here discussed a peculiar relation obtains. The large central areas of granite, whether sheared, as along the eastern margin of the Bitterroot Mountains, or massive, as usually is the case, seem conspicuously barren of deposits. The vein systems appear in or close to the four smaller areas of sedimentary or metamorphic rocks which are found at the periphery of the great central granite area. It is thus in the quartzitic series of Lolo Fork, in the quartzites, slates, and gneisses of the upper South Fork of the Bitterroot, and in the old gneiss areas of Elk City and Pierce. While the age of the quartz veins is not established beyond doubt, it is probable that they were formed during the latter part of the Mesozoic era.

---

[a] The gold and silver veins of Silver City, De Lamar, and other mining districts in Idaho: Twentieth Ann. Rept. U. S. Geol. Survey, pt. 3, 1900, p. 238.
[b] Ibid., p. 238.

NORTH END OF BITTERROOT RANGE.

Lolo Fork.—The gravels of Lolo Fork are said to contain a little placer gold, although there are no evidences of extensive workings. Quartz veins containing gold and copper are found in the calcareous quartzites 5 miles above the mouth of the river. One of these, called the Chickerman vein, crops half a mile below Westerman's ranch. The quartz contains a little chalcopyrite and an 8-stamp mill was erected on the property in 1889. There is no evidence, however, of long-continued working. One mile below Westerman's, on the south side of the river, is another quartz vein containing copper minerals and referred to as the Sherman and Lattimer claim.

Some copper prospects are located 9 miles above the mouth of the river and 2½ miles above Westerman's ranch. The best prospects are contained in the limestone in Woodmans Creek, a tributary entering Lolo Fork from the north. A quarter of a mile from the river a vein crosses the gulch with a northeasterly strike. The decomposed croppings were worked in arrastre many years ago, the shallow oxidized zone having produced about $5,000. The tunnels driven in from Woodmans Creek have developed some good-looking copper ore, chiefly containing chalcopyrite and bornite. Five carloads of ore are said to have been shipped. About a thousand feet above this vein a dike of dioritic rock 10 feet wide appears in the bottom of the canyon. This rock contains chalcopyrite and specularite scattered through its mass.

Mormon and Carlton creeks.—Several quartz prospects have been located on Mormon Creek, which heads near Lolo Creek and empties near the mouth of Lolo Fork. There is a considerable evidence of mineralization in this vicinity, and the zone is probably the continuation of the vein systems described from the lower Lolo Fork.

One mile north of Carlton Creek the quartzites bear evidence of some mineralization along the foothills of the Bitterroot Range, and this again may be considered to be a general continuation of the mineral belt referred to above.

Few other prospects are known in this vicinity. It is stated, however, that galena veins have been found and prospected high up in the mountains on Bass Creek, a short distance north of St. Marys Peak.

Curlew mine.—This deposit, which has produced a considerable amount of silver-bearing lead ore, is situated in the first foothills of the Bitterroot Range, at an elevation of 3,533 feet, 14 miles north of Hamilton and just south of the mouth of Big Creek Canyon. The mine is situated at the very point where the first foothills emerge from the Pleistocene of the valley. For a distance of half a mile west of the mine the rocks consist chiefly of limestones and quartzites, both usually much shattered and without clearly defined bedding. In places the quartzites grade over into typical contact metamorphic hornfels. The normal gneiss of the Bitterroot Mountains adjoins the quartzite and is well exposed along the canyon of Big Creek, the dip being, as usual, 25° E.

The deposit was not discovered until 1888, and was worked from that time until 1893. The total production is reported to have been $800,000. The principal developments consist of a perpendicular shaft 500 feet deep with crosscuts on several levels to the vein. The deposit shows few distinct outcrops, but as developed by mine operations strikes north and south and dips 45° E. The workings are naturally not accessible at the present time, the shafts being filled with water, and the information here given was communicated to me by the former foreman of the mine, Mr. Thomas Corvan. The ore principally consists of galena; the first-class ore contained 55 per cent lead, 125 ounces of silver, and from $2 to $7 gold per ton. The ore body was found along a well-defined fissure on an average 10 feet wide and was all along on the east side protected by a thick water-tight cover of clay. Drifts penetrating this clay broke into gravel and sand containing much carbonized wood.

From this information, which is doubtless correct, it would appear that the vein occupies a fault fissure between the quartzite and the Pleistocene gravels of the valley. This, besides being most unusual from a mining standpoint, has a structural significance which has been further discussed on page 48.

A few hundred feet west of the shaft is a prominent outcrop showing traces of copper, lead, silver, and gold. This is probably part of a vein which continues southward for half a mile to the Whippoorwill claim; it is in badly crushed lime and quartzite the entire distance. Northward this second vein continues to the east Curlew shaft where it was mined by the Curlew Company and found to contain a rich body of galena ore; $200,000 are said to have been extracted from this shaft.

The Pleasant View prospects.—Several quartz veins containing silver have been prospected in the first foothills of the Bitterroot Range about 10 miles north-northwest of Hamilton and at an elevation of about 3,900 feet. The country rock is a granite forming a small area in front of the usual gneiss of the Bitterroot Range. The prospects are not worked at the present time and the developments are not extensive, chiefly consisting of two or three shafts of a maximum depth of 100 feet. There is one especially strong quartz vein with a northeasterly direction. The ore consists of quartz containing a little pyrite and tetrahedrite. The values are chiefly in silver.

UPPER SOUTH FORK OF THE BITTERROOT.

Trapper Creek.—No deposits are known to occur south of the locality just described until the southern end of the range is reached; no placers have been worked and the creeks appear to be barren of even traces of gold. A quartz vein was noted on Trapper Creek, near the south end of the range and about 3½ miles from the mouth of the canyon. This is a shear zone 6 to 8 feet wide, occurring in granite, and containing several stringers of quartz. The strike is north-south and the vein stands nearly vertical. An assay was made of the quartz from this prospect, the result showing 0.06 ounce of gold and 0.30 ounce of silver per ton. No deposits of any kind have been reported from the Nez Perce Pass, or the creek leading up to the pass.

Slate Creek.—About 18 miles south of the forks of the river Slate Creek joins it from the east. On this tributary are located several veins which were visited by Mr. G. W. Stose. The elevation at the mouth of Slate Creek is 4,700 feet. Near the entrance to the gulch are four or five cuts on the north slope on quartz veins containing calcite and tetrahedrite. These prospects are contained in granite-porphyry. On Slate Creek, several miles above its mouth, a well-defined vein crosses it with a direction a little west of north. On this vein, which cuts through granite, quartzite, and porphyry several locations have been made. The one next the creek is called the Moon mine, and is continued southward into the "Tiptop." At this place is exposed 6 feet of quartz dipping steeply to the west. The ore contains bornite, chalcopyrite, and tetrahedrite. The third location visited is "Dynamo" which is developed by a shaft 35 feet deep. Other prospects are located a short distance north of the Dynamo and among them is mentioned the Empire and the Carney. These are said to contain galena and chalcopyrite. It is not known that any of the prospects on Slate Creek have any considerable production to their credit.

Hughes Creek.—Eight miles south of Slate Creek, and about 10 miles north of the Salmon River divide, Hughes Creek joins Bitterroot River from the east. At its mouth is located the post-office of Alta (elevation 5,160 feet). The rock exposed here and for long distances above on the gulch is a heavy-bedded quartzite. For about 10 miles above the mouth the creek has a rather wide bottom with occasional meadows. In places the width of the bottom lands is a quarter of a mile. The walls of the canyon ascend with steep slope to a thousand feet or more above the bottom. Hughes Creek is known since the early days of mining as gold-bearing, and a continuous string of claims cover the bottom from the mouth up to the divide, 15 miles distant. For 10 miles above the mouth the gravels are deep, and bed rock has not been reached. At present the principal operations are carried on in the upper part of the creek.

The gold is generally coarse and occurs in flat grains accompanied by garnets and black sand. All along, gravel terraces accompany the watercourse, although in many places they have been removed by erosion. Near the mouth there is a gravel bar 50 feet above the stream, and another one 200 feet above the same. Evidences of still higher gravel bars are found up to an elevation of 500 feet above the bottom, and many claims are located on these high bars. Whether any of these occupying larger areas near the mouth of the creek will pay, for working is not certain.

In 1899 at least three companies were working the creek, the first consisting of Orton, Molloy & Co. These men were ground sluicing—shoveling the material into the sluice boxes. A little higher up are the Keating claims. One mile above this the Wood Placer Company was beginning operations on a larger scale, using an elevator. In 1902 it was reported that the Wood Placer mines had been purchased by eastern capital and that a 4-mile ditch was being constructed which will furnish a pressure of 300 feet. The company expects to operate two giants. Stream tin is frequently found with the gold in this creek.

Johnson Creek, a few miles above Hughes Creek, is also said to contain a little gold.

Mineral Point.—This mining district is located about 6 miles north of the Salmon River divide, near the head of Bitterroot River, at the point where the stream splits up in several branches. The elevation is about 5,560 feet. The formation consists of quartzitic and chloritic schists. A number of copper-bearing veins have been located in this vicinity, and some small shipments are said to have been made. The first prospects are located on Woods Creek, about 1 mile west of the main river. The croppings are found near the contact of the quartzite with gneiss or pegmatitic granite. The developments are very slight. The rusty, decomposed ore shows pyrite, limonite, and azurite. The principal prospect in the Mineral Point district is said to be the Copper Queen, located near the road at the upper end of the meadows on the main fork. The ores are principally chalcopyrite, containing a little gold and silver; none of the gold is free. The showings on the surface were good, but a 400-foot tunnel run 100 feet below the croppings only developed schists impregnated with pyrite containing a little gold and copper. The best ore of the Copper Queen contained, it is said, from 5 to 20 per cent of copper and 11 ounces of silver, with a few dollars in gold per ton. Many other irregular veins of quartz stained with copper and showing evidence of considerable pressure and deformation were noted in the vicinity.

Up toward the adjacent district of Mineral Hill are several veins in granite or gneiss containing galena and arsenopyrite with from 6 to 24 ounces of silver per ton. The prevailing strike is northeast and southwest, while the veins generally dip 70° S.

Mineral Hill.—This district adjoins Mineral Point on the south and is chiefly located on the upper slopes of the hills which descend toward Salmon River. South of it, and almost connecting with it, is the gold-mining district of Shoup, in the canyon of the Salmon River. This, however, was not visited.

The veins of Mineral Hill are chiefly located in the upper part of Owl Creek, on the ridge south of Blue Nose Mountain, and also in the upper part of Horse Creek. The veins occur in a coarse gneiss cut by many dikes of granite. The only prospect worked in 1899 was the Monitor, located at the head of Owl Creek, at an elevation of 7,500 feet. A road was being built about this place from the headwaters of Bitterroot River. The Monitor is a 5-foot quartz vein with an east-west strike and containing scattered bunches of galena. A shaft was sunk on it to a depth of 50 feet and a considerable amount of ore has been shipped. It contained lead, with up to 100 ounces of silver, and a little gold.

Near the head of Horse Creek, at an elevation of 6,900 feet, are several veins striking east and west and dipping 40° N. The quartz contains a little galena and pyrite and is slightly copper stained. Several tunnels have been run on these veins which, however, at the time of visit, appeared to be abandoned.

GIBBONSVILLE DISTRICT.

The Gibbonsville mining district is located in Idaho, on a tributary of Salmon River, and only about 10 miles distant in a straight line east of the Mineral Hill district just described. Gold-bearing quartz veins are found near Gibbonsville, and have been worked for many years with varying success. Some years ago the American Development and Mining Company, who owned the principal mines in the district, introduced barrel chlorination to treat the ores. This was not found to be in all respects satisfactory, and lately a cyanide process has taken its place. The district was visited by Mr. W. H. Weed in 1898, who kindly put the following notes at my disposal: The rocks at Gibbonsville are steely gray, quartzose, and micaceous slates, thinly laminated. The slates strike north and south, and dip from 35° to 65° to the east or to the west. The quartz veins run nearly east and west, sometimes southeast and northwest, while the dip is northerly. The main tunnel by the American Development and Mining Company cuts at least six veins, with sharp, smooth walls, varying in thickness from a few inches up to 15 inches. The quartz carries abundant pyrite, and contains from $30 to $50 in gold, together with about $1 in silver. The oxidized part of the veins is free-milling, while below that zone very little native gold appears. Occasionally the quartz veins are banded or ribboned by parallel fracture planes. While the veins are very persistent, they are not uniformly ore-bearing, and no absolute regularity of the distribution of the ore has been observed.

ELK CITY DISTRICT.

PLACER MINES.

Important placer mines have been worked for many years on the upper tributaries of the South Fork of the Clearwater, this whole producing region being referred to as the "Elk City district." Placers occur on Red River and its numerous tributaries, on Red Horse Creek, American Creek, and Elk Creek; finally, on Crooked River, Moose Creek, and Newsome Creek. Below this point the gravel bars of the South Fork have been mined for many miles—at least, in places— as far down as Stuart, at its junction with the main fork of the Clearwater. The Elk City diggings were discovered in 1861. During the following years the activity was great, and the production doubtless amounted to several million dollars. The gravels in the bottoms of the streams were generally too low to be easily reached, and most of the operations were confined to the high bars and gravel terraces. In many places in the vicinity of Elk City the so-called "skim" diggings yielded largely, this class of deposits embracing the shallow concentrations of placer gold over large areas within a foot or two of the surface. For many years after the first few seasons of heavy production the camp was turned over to the Chinese. Renewed interests in the placers were, however, taken by white men during the last ten years, and in 1899 several deposits were actively worked on a fairly large scale.

The production of this district has been referred to on page 84. Suffice it to say that since 1875 it has yielded something like $30,000 to $70,000 per annum. As there is a fairly large amount of workable deposits remaining, a moderate output will probably be continued for many years; very likely, however, with a decreasing tendency. A map on a larger scale is shown in fig. 7.

Red River.—Coming down into Red River from the Nez Perce trail much quartz-gravel is seen in the main stream and its tributaries. In many of the latter are abundant evidences of old and extensive placer mining. At French Creek, a tributary entering Red River from the north a short distance below the post-office of Raymond, and at an elevation of about 4,500 feet, a dredger was working the old tailings and the gravels underneath them. This dredger is owned by the same company which was operating the hydraulic mine at American Hill.

Along Red River, bottom lands and meadows extend from Ryan's ranch for 6 miles down. Below this point the river flows in a narrow canyon, opening up again near the mouth of American River and Elk Creek. Many placer mines are located along the gulch gravels of Red Horse Creek, which heads near Anderson Butte and empties at the lower end of the Red River meadows. Within this upper basin just described there are few evidences of terraces or high gravels, but a different condition of affairs appears as soon as the basin of Elk Creek and the American River is entered.

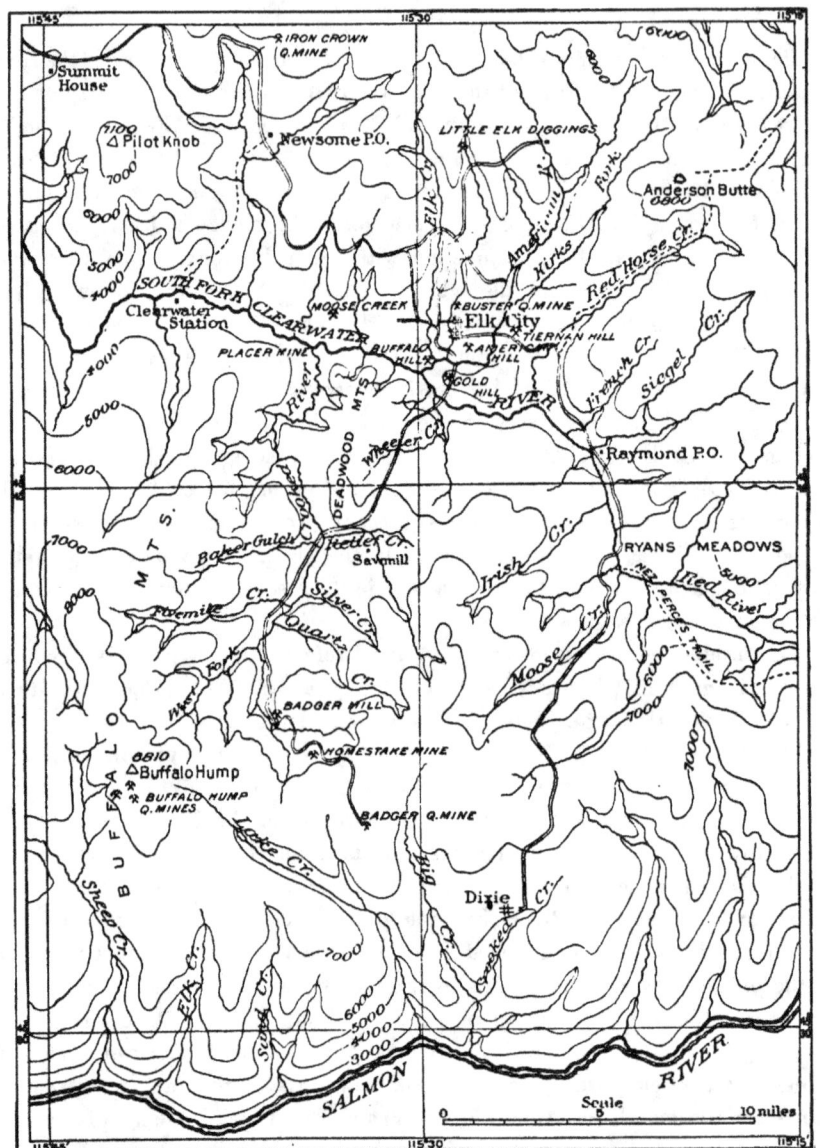

Fig. 7—Map of upper South Fork of Clearwater River, showing location of mining districts.

The basin of Elk Creek and American River is occupied by a system of gravel benches filling the basin to an elevation of somewhat over 4,500 feet.

From any prominent point the level of this gravel terrace is seen to skirt the valley for a long distance westward, and the higher mountains of Newsome Creek rise with more decided slope from this general level of deposition. For 2 or 3 miles above its mouth Elk Creek, emptying into the South Fork of the Clearwater from the north, contains broad bottom lands and meadows. A first bench is marked fairly clearly 50 feet above the creek and is covered with well-washed quartz gravel. This bench is well exposed at the little settlement of Elk City. On Little Elk Creek, about 3 miles above its mouth, the bottom lands narrow, and gravel ridges 100 feet high rise on both sides of the stream. At about 325 feet above Elk City, or an elevation of 4,525 feet, the above-mentioned prominent level of deposition is reached. This terrace of well-washed gravel is seen to skirt the ridges separating Elk Creek from Newsome Creek and continues across the divide between Elk Creek and the headwaters of American River. Remains of this same gravel terrace are also found in the Crooked River basin, although here a more active erosion has removed a larger part of these old deposits.

American Hill.—The most prominent gravel mine worked in 1899 is located at American Hill 1 mile southeast of Elk City, at an elevation of 4,200 feet, the elevation at Elk City in the upper part of the town being 4,200 feet approximately. The deposit at American Hill is owned by the Idaho Mining Company. About thirty years ago a little work was done here, chiefly in the nature of "skim" diggings and some drifting. The present company eight years ago removed the gravel down to the bed rock and exposed an old channel of the American River running from northeast to southwest. One thousand miner's inches of water is used, and an elevator raises the gravel to a series of sluices aggregating 500 feet in length and having a grade of 5 inches in 12 feet. Aneroid measurements gave the approximate elevation of the bed rock as 4,165 feet, or about equal to the lower portion of Elk City, and not much higher than the bed of American River a short distance south of the mine. The bed rock consists of a soft, coarse gneiss intruded by abundant pegmatitic veins. The gold is fairly coarse, much of it being of the size of wheat grains. The slope of the bed rock is said to be from 3° to 4° W.; the banks exposed in the pit, which has an area of about 700 by 200 feet, are from 60 to 100 feet high; the material exposed is prevailingly of a sandy character.

The top stratum consists of 10 feet of gravel and clay. This stratum was rich and worked in the early days as "skim" diggings. Below this follow 50 feet of sand, clay, and gravel interstratified. The lower portion of the deposit consists of 20 feet of red gravel, underneath which is found 15 to 20 feet of bluish gravel which contains the principal quantity of gold. The gravel is mostly quartzose, and the large bowlders on the bed rock often contain native coarse gold. On the south side of the pit the workings have exposed a decided bluff of bed rock 30 feet above the general level of the bottom of the channel. Not unlikely this represents a fault since the deposition of the gravel.

Tiernan Diggings.—A few miles above American Hill on the river of the same name are located some high gravels of considerable interest. The place is close to the road from Elk City to Raymond. The gravels are considerably higher than those at American Hill and they probably do not belong to the same system of channels. Going east from Elk City a high ridge of gneiss is crossed. Descending again to American River a distinct bench covered with wash gravel is noted at an elevation of 4,300 feet. At the bridge the elevation is about 150 feet less and the exposures consist chiefly of gneiss.

Tiernan Hill is situated on the east side of the river, a short distance from the bridge. The bed rock has an elevation of 4,450 feet and the top of the bank 4,580 feet. On the uneven bed rock lie thus 130 feet of sediments, chiefly consisting of sand and clay with some interstratified gravels. In former years a considerable amount of hydraulic work was done at this locality and the amount of remaining gravel is probably not large. In elevation, Tiernan Hill corresponds to the high terrace skirting the Elk Creek basin, but is, if anything, somewhat higher than this. It has the appearance as if this deposit formed a part of a larger channel, the central bed of which has been removed by erosion. A little work is still done each year.

Buffalo Hill.—The gravel mine of this name is situated on the north side of American River about a mile and a half southwest of Elk City, and near the point where Elk Creek empties into that stream. Something like 180 acres of gravel are contained in this isolated area. The larger part of this has already been mined by hydraulic operations, though a considerable quantity of poorer gravel still remains. Active work was in progress during 1899. The available fall is slight, hence there is some difficulty in disposing of the tailings. The flume is 2,000 feet long and has a grade of 3 inches in 12 feet, wooden riffle blocks being used. The size of the flume is 4 by 4 feet. The principal operations have been carried on on the north side of the hill, where several acres of bed rock are exposed. In front of this lies a considerable mass of tailings.

The bed rock is throughout a soft, coarse gneiss of the kind so prevalent about Elk City. The configuration of the bed rock shows plainly that the deposit is a part of an old channel of the American River. A high rim lies immediately north of that stream and is 150 feet above it. North from this rim the bed rock descends until at the principal pit the elevation is about equal to that of the bed of the American River. According to aneroid measurements, the elevation is 4,150 feet, or about 50 feet lower than the average elevation of Elk City.

The exposed banks show 40 feet of prevailing sand with some gravel. A somewhat coarse gravel rests immediately on the bed rock and consists chiefly of cobbles of quartz with some gneiss and also a large amount of black quartzite; the latter is probably derived from an area exposed on Relief Creek a few miles southward.

The gold is both of a fine and coarse variety, hence the necessity for long sluices. It is as usual principally scattered on the bed rock.

North of Buffalo Hill, on Buffalo and Blackfoot gulches, the ordinary bed rock is exposed, although still higher up gravel of the upper terraces is encountered. A very interesting feature is the exposure in the hydraulic pit of the mine of a few acres of so-called "false" bed rock. This consists of a series of heavy-bedded clays containing much coaly material, and is evidently a part of a sedimentary series resembling lake beds which prior to the accumulation of the gravels covered a part of this country. On the northwest the clays border against gneiss, the separating plane being a well-defined perpendicular fault line, the downthrow being on the southeast side.

A mile to the southeast of Buffalo Hill is Gold Hill, situated on the point between Red River and American River. The elevation of the bed rock on the south side of the road is here 4,300 feet; these gravels are thus 150 feet higher than those of Buffalo Hill. The thickness of the gravel at this point does not seem to be considerable. Washing on a small scale is carried on at this point each year.

South of Red River the road from Elk City to Crooked River ascends Wheeler Creek, along which shallow diggings are found at intervals at elevations from 4,200 to 4,250 feet. The flat bottom of this stream has been ground-sluiced all the way up to the divide toward Crooked River, which has an elevation of 4,900 feet. Much of the gravel is of a subangular character. A few well-washed quartz pebbles were noted almost up to the divide.

Crooked River.—The road from Elk City to the head of Crooked River descends to the latter by way of Relief Creek. In the flat bottom of this, evidence of old placer diggings are found. Quartzite cobbles are very abundant, but no great amount of bench gravels were seen. Crooked River and the gulches emptying into it have been worked for several miles above this point. The gravel chiefly consists of the present accumulations in the creek, together with smaller amounts of low bench gravels. The mouth of Crooked River was not visited, but it is reported that the Tacoma and Clearwater Mining Company was operating a hydraulic mine near this place.

Little Elk Creek.—Three miles north of Elk City the Little Elk Mining Company was preparing to wash the shallow gravels spread out over the bottom of this creek. A short distance north of this point the actual gneiss bed rock appears in the creek bottom, but the material on which the gravel rests at this mine is what is usually termed "false" bed rock. This consists here of clays and micaceous sands with a considerable amount of coaly material. It is exposed at two places below the pay gravel and has a dip of 60° E. These beds have very much the appearance of being deposited in a broad river or lake and present great similarity to the beds of the same kind observed at Buffalo Hill.

Newsome Creek basin.—About 6 miles due west of Elk City the Moose Creek diggings are located on the ridge dividing the creek of the same name from Newsome. The place was not visited, but from accounts the deposit appears to be of the same kind as that exposed at Buffalo Hill. The banks are said to be 160 feet in height, the gold, both coarse and fine, varying from $16 to $19.20 per ounce. This mine was operated in 1899.

From Buffalo Hill the road to Harpster, on the Clearwater River, ascends over a long ridge of wash gravel. The top of the terrace, which is the same as observed on the headwaters of American River, has an elevation of about 4,450 feet. Scattered wash occurs above this point up to an elevation of 4,750 feet. Above this point the gneiss rises in low ridges. The same gravel terrace continues into the drainage of Newsome Creek. This stream has been washed all along its course, as have also the benches which accompany it at an elevation of 30 feet above the water. The bottom is generally flat and filled with tailings to a width of 100 or 200 feet. An important gravel terrace appears 300 feet above the creek and is well exposed both on the east and west side of Newsome post-office. The elevation of the bench is 4,400 feet. These high bench gravels have not been generally washed, though placer operations are frequently carried on in the small gulches which traverse the terrace and in which a secondary concentration has taken place.

Along the South Fork of the Clearwater the low bars have been mined for many miles downstream; at least as far down as Stuart. High gravels corresponding to those of American River and Newsome Creek were not observed. There are, however, occasional small gravel benches 200 feet above the river. In many cases these have also been washed. The present production from this kind of mining is doubtless very small.

QUARTZ VEINS.

For many miles west of the mining districts at the head of the Bitterroot River there are no indications at all of mineral deposits until the drainage of the Red River is entered, this being a tributary to the South Fork of the Clearwater River. Near the head of Red River, on the Nez Perce trail, between Ten Dollar and Mountain Meadows, the first indications of quartz veins are found, and a few prospect holes have been opened on the ridges, which are covered by a deep soil, the prevailing rock appearing to be granite. Descending into the Red River, prospects are again found on the ridge north of Ryans Meadows. From here on to Elk City, a distance of 10 miles, all creeks show evidence of placer-mining operations, and occasional quartz prospects are noted.

While the gravels near Elk City frequently contain quartz bowlders with native gold, indicating the abundant presence of auriferous veins, very few of these veins have been actually found and worked; prospects are known from several places, among others on Red Horse Creek, but the developments are in all cases slight. The only quartz vein prospected during 1899 in this vicinity appears to be the Buster mine, which is located three-fourths of a mile north of Elk City, on the east side of Elk Creek. The strike is easterly and westerly, and the dip appears to be 65° S. The deposit is a typical quartz vein of moderate width, containing pyrite, zinc blende, galena and tetrahedrite. The quartz is coarse and shows normal vein structure. Many long, slender crystals are embedded in a mass of smaller individuals. The surrounding rock is gneiss, and the vein cuts across the schistosity. An assay of selected ore, without visible gold, gave 3 ounces of gold and 6½ ounces of silver per ton. No ore was being extracted during the time of the visit, and the developments are not extensive. Since 1899 many new locations and generally increased activity in quartz mining are reported.

The Newsome Creek basin contains important placer diggings, and many prospects have been found in the surrounding hills. None of them, however, have been extensively worked. A couple of miles north of Newsome post-office a small vein of normal comb quartz was found in gneiss, having a strike of N. 30° E., the dip being 20° to the N. This prospect is called the "Lookout" mine. A little higher up, in a gulch leading up from the east side of Newsome Creek, is the Iron Crown vein. The elevation is here approximately 4,700 feet. A mill has been erected, and is shown in Pl. XV, B. Both mine and mill were being operated on a moderate scale at the time of visit, and a considerable amount of ore was produced every month. The developments were not very extensive, the deposit having the peculiar form of a blanket vein of rusty, decomposed quartz from 2 to 6 feet wide and situated at a depth of only a few feet from the surface. The gold is free, which might be expected from the oxidized character of the ore. The inclosing rock is the gneiss typical of this section. This mine was the only producing quartz mine in the Elk City and Newsome Creek districts during 1899.

Throughout the drainage of Crooked River, which enters the South Fork a short distance above Newsome Creek, abundant placer diggings occur, but strangely enough very few quartz veins of value have thus far been encountered. Near the head of Crooked River several veins have, however, been located.

The principal producing mine is situated near the head of the river and practically on the divide separating this drainage from the Salmon River. It is called the Badger mine and a few years ago was the scene of active mining operations. A 10-stamp mill was built on the Crooked River about 5 miles northeast of Buffalo Hump and a road constructed from there to the mine. The vein is said to occur in granite and is a normal quartz vein containing a considerable amount of free gold.

Another vein, called the "Homestake," is situated on the road to the Badger mine, 1,300 feet above the mill and at an elevation of about 6,400 feet. The Homestake crops in granite and has a strike of N. 80° W., the dip being 70° N. The vein appears to be only from 2 to 3 feet wide and is developed by two tunnels 100 feet apart. The ore is a white normal quartz containing pyrite and a little galena.

During 1902 the Crooked River Mining and Milling Company is reported to have completed a 20-stamp mill at Oro Grande, on the Crooked River, 12 miles in a southerly direction from Elk City and three-fourths of a mile below the old Badger town and the forks of Crooked River. The deposit which has been developed during 1902 is said to be a very wide mineralized belt of low-grade free-milling ore, striking north and south. The mining and milling expenses are claimed to be less than 50 cents per ton. The same company owns the Orion and the Regnier quartz veins in the same district. New discoveries of veins are also reported 3 miles below Oro Grande and north of Deadwood Mountain and Relief Creek.

In 1901 the following companies have reported production of gold and silver from the Elk City district: Gold Hill Placer Company, Little Elk Mining Company, Tiernan placers, Relief Creek placers, American Hill placers.

DIXIE DISTRICT.

A mining camp which is of small extent, but which has maintained a certain production for many years, is located at Dixie, on the headwaters of Crooked Creek, which drains into Salmon River almost due south of the point where the larger stream of the same name enters the Clearwater River. The locality was not visited. A wagon road extends to it from Elk City, following up Red River to near Ryans Meadow, and continuing thence southward. The placers of Dixie camp consist partly of the gravels in the creek beds and to some extent of higher benches. Cinnabar is said to have been found in the gravels, with the gold. Numerous smaller gold-quartz veins are reported to exist near the camp, and have been worked at intervals on a small scale. During 1899 considerable prospecting was in progress in the country between Dixie and the Badger mine, about 8 miles farther west, and promising veins are reported to have been uncovered.

BUFFALO HUMP DISTRICT.

The divide between the Salmon and the South Fork of the Clearwater, in the vicinity of Dixie, is a broad, flat ridge rarely rising to elevations of 7,000 feet. About 10 miles west of this vicinity it rises to a prominent bare peak, attaining a height of 8,800 feet above the sea, which is commonly referred to as Buffalo

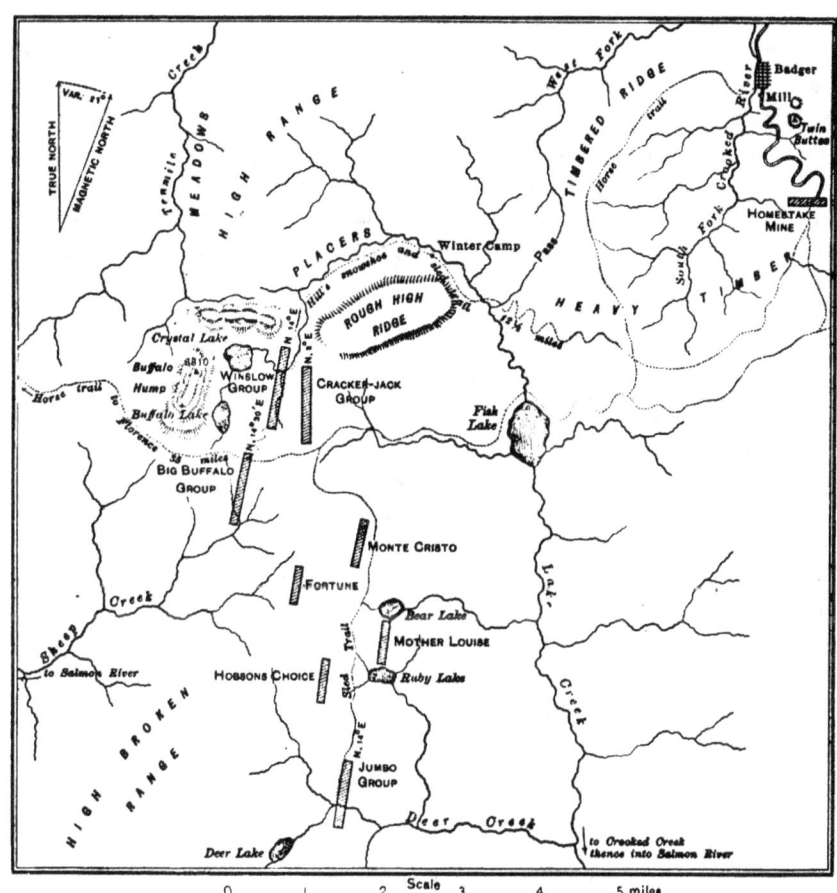

FIG. 8.—Sketch map of the Buffalo Hump mining region.

Hump. Quartz veins of some importance were discovered at this vicinity in 1898. The Robbins mining district was organized to cover these locations, and in 1898 and 1899 active prospecting was progressing about that camp. The principal veins are situated on the south side of the peak on a broad plateau sloping gently westward and having an elevation of about 8,000 feet. In 1899 the place was only accessible by means of trails, one leading up from Crooked River and another one, more frequently used, from Grangeville. The latter route followed the Florence road about 20 miles southward to Adams camp, from which place a trail continued to Buffalo Hump, leading over high and bare ridges. It is reported that shortly after that date a wagon road was constructed, following the last-named trail. The general drainage and the location of principal claims is illustrated on fig. 8, reduced from a map by Mr. W. H. Hill.

From the Badger mill located on Crooked River, the trail ascends westward over sloping ridges to a broad divide covered with thick forest. From this point it descends into Lake Creek, a long tributary to the Salmon River, heading on the east side of Buffalo Hump. When Lake Creek is reached the trail turns up the same, following it over heavy moraines, ascending about 3,000 feet. The divide at the head of Lake Creek has an elevation of nearly 8,000 feet. West of it, and at the foot of Buffalo Hump, extends a flat, marshy in places, and containing several small lakes or ponds.

The predominating rock at Buffalo Hump is the ordinary Clearwater granite of light-gray color, and containing both biotite and muscovite. It is on the whole very similar to the large areas of the same rock which extend to the south of Salmon River in the vicinity of Warren camp. Near Buffalo Hump this granite contains a number of irregular patches and smaller areas of micaceous schist, quartzites, and coarse granular marbles, the latter very largely converted into greenish and brownish hornfels and garnetiferous rocks. The strike and dip of this sedimentary series are extremely variable, but are usually at moderate angles of 30° to 40° to the east or north. It is evident that this sedimentary series has been intruded by the granite and suffered intense contact metamorphism. The schists are penetrated by granitic dikes, running in all directions. Regarding the age of the schists, nothing more definite can be said than that they are older than the intruding granite. No other sedimentary areas are known in the vicinity with which they can be even tentatively connected. The granite often shows two well-marked systems of joint planes, one having a northeasterly and the other a northwesterly direction. Mr. Charles L. Whittle,[a] who has published a short but interesting article on the district, states that "along the northeasterly set of joints, and to a less extent along the northwest set, there have been introduced small quartz veins in great abundance. These veins are the oldest in the camp and vary in width from mere films up to several feet, the great majority being about 1 inch across. These small veins seldom appear mineralized and usually lack permanence, either horizontally or vertically." The producing veins form a linked system diverging slightly northward and having

a Eng. and Min. Jour., vol. 68, 1899, pp. 215, 216.

a strike of a few degrees east of north, the dip being generally very steep toward the east. The first vein discovered was called the "Big Buffalo." This claim is situated on the high plateau already referred to. A 10 by 12 foot shaft was sunk to a depth of 72 feet in 1899 and the drifts and crosscuts amounted to 100 feet. The dump covered a large part of the outcrop, in which the vein was stated to have a width of 40 feet. In a cut just north of the shaft about 12 feet of vein quartz were exposed. The ore is a normal, milky-white quartz containing scattered masses of pyrite, galena, copper pyrite and tetrahedrite. An assay of some of the material from the dump gave the high result of nearly 8 ounces of gold and 7 ounces of silver per ton. It is stated that the Big Buffalo vein can be traced southward through three claims and is then lost along the line of the east and west valley of Sheep Creek, which drains the high plateau in a westerly direction and finally turns southward toward the canyon of the Salmon River.

A few claims south of the Big Buffalo is the Fortune vein. This was found to have a strike of N. 10° E. and a steep easterly dip. Two feet of quartz showed between well-defined walls and contained scattered crystals of pyrite.

A short distance east of the Big Buffalo another prominent vein appears and is continuous for a distance of 1 mile. Near the north end of this is the Monte Cristo claim, which in 1899 was developed by several small tunnels. The vein appears between granite walls. This is continued southward by the St. Louis and other claims, showing again a well-defined quartz vein from 3 to 6 feet wide. The normal vein quartz contains small grains of pyrite, chalcopyrite, galena, and zinc blende. An assay of a sample taken from a prospect hole 10 feet deep on the St. Louis gave 3.4 ounces of gold and 9.2 ounces of silver per ton. Still farther south on what appears to be the same vein is located the Jumbo. On this claim the deposit has been opened by two tunnels, the lower one 300 feet in length. The vein appears of fair size and of a fairly uniform thickness for several feet. Here again the quartz contains a little iron pyrite, tetrahedrite, and chalcopyrite.

On the eastern side of the camp, at the head of Lake Creek, a great number of prospects were located, the most important of which were called the Vesuvius, Crackerjack, and Winslow. On all of these active work was in progress and fair ore was exposed. Owing to the glaciated character of the country, there is but little decomposed vein matter with free gold. Ordinarily shallow prospect holes will reach fresh sulphides. It is stated that the quartz is free milling, although it seems probable that the larger part of the values are contained in the sulphurets.

A number of placer claims have been located in the vicinity, but owing to the extensive glaciation it is probable that most of these will be disappointing in yield.

According to late information from Mr. V. C. Heikes, of the division of mining and mineral resources of the United States Geological Survey, the following properties were producing at Buffalo Hump in 1902: The Jumbo has been working 4 stamps during the year, and unverified reports placed the yield at $5,000 per month. On the Crackerjack a 5-stamp mill was running in September, and during that month treated about 400 tons of ore. In addition to concentrates and tailings the clean up from the plates for the month is given as $4,000.

On the Wise Boy a mill had been erected and was preparing to start on October 1. This claim is believed to contain ore of excellent quality.

The Buffalo Hump syndicate has been working 25 men, and the 10-stamp mill on the Big Buffalo has been operated steadily for the last ten months. Bullion has been shipped regularly.

In the article quoted above, Mr. Whittle expresses himself as follows:

"I have been greatly disappointed to note the scarcity of free gold in the oxidized portions of the veins, and the sudden and universal thinning out of the prominent veins, their unequal pockety mineralization, and the strong tendency of the veins to pass from well-defined walls into a ramifying mass of small stringers within a short distance has filled me with apprehensions as to the future of the camp."

Judgment of the real character of the veins would, however, best be deferred until the ore bodies shall have been thoroughly exploited. The result of the last two years' work shows that to some extent Mr. Whittle's predictions had a foundation. At the same time it is evident that some shoots of value have been encountered and that the camp will continue, for a long time, as a producer, though not on as large a scale as was predicted during the first boom.

PIERCE DISTRICT.

PLACERS.

The Pierce district was discovered in 1860, and for the first few years after that date a large production was maintained. Soon, however, most of the mining population was attracted to richer camps elsewhere and the production declined greatly. For the last ten years it has probably rarely exceeded $20,000 or $30,000 per annum. As in so many other cases, the total production is impossible to obtain. Probably the production was considerably less than at Elk City, which, in turn, was eclipsed by Florence, Warren, and the Idaho Basin. A guess may be hazarded that the total output of Pierce is in the vicinity of $5,000,000. A certain amount of placer mining is still done every year, and at intervals some of the numerous quartz veins are also operated. Pierce is situated near the head of Oro Fino Creek, which empties into the main Clearwater River 25 miles farther west. The elevation of the camp is in the vicinity of 3,000 feet. From the monotonous lava plateau extending about Weippe a wagon road leads to Pierce, a distance of about 12 miles. After a few miles of continued Columbia River lava, low hills of granite or imperfectly gneissoid rock rise, and over the same rocks the road finally descends to Oro Fino Creek. Pierce appears to be located at the contact of an extensive area of gneiss and mica-schist, which continues toward the northwest across the Mascot Hills to the North Fork of the Clearwater. South of the contact granitic and dioritic rocks prevail. Near the contact the schists are penetrated by granitic dikes. Below Pierce, Oro Fino Creek runs in a narrow canyon, which widens and splits up into two branches close to the settlement. The gravels worked are partly the low benches along the creek, partly also high terraces extending to a height of 500 feet above the stream. Canal Gulch extends northward from Pierce for several miles, heading at the divide toward the North Fork of the Clearwater. It has been extensively worked, and at least in one place a half-mile north of town a high gravel bench is still mined. At this place there appear to be two levels of terraces, one 100 feet above the creek, while the bed rock of the other one is 200 feet above the same level.

The bed rock is a coarse diorite. The banks are 50 feet high and the gravel contains a great abundance of quartz pebbles.

Above Pierce, Oro Fino Creek has been worked for several miles. On the road to Rhodes Gulch well-washed gravel extends to a height of 500 feet above the town. Rhodes Gulch, coming down from the north and joining the main Oro Fino Creek a few miles above the town, is also reported to have been very rich, which is also confirmed by the extensive dumps of washed gravel covering its bottom. East of this point is French Creek Mountain, and a road leads across to the quartz prospects which have been found there.

Altogether Oro Fino Creek has been washed for gold for a distance of 12 miles above and 10 miles below Pierce. The gold is usually fine as to size, but of low grade as to value, being, at most, worth $17 per ounce, some of it running as low as $15.

A few miles below Oro Fino, not far from the tributary called Poorman Creek, the American placer mine was in operation in 1899, washing the gravels of the creek bottom. The creek at this point flows over a basaltic bed rock.

QUARTZ VEINS.

As may be expected a number of quartz veins have been found and located near Pierce. None of them has, however, thus far reached the standing of large mines. As so often happens in districts of this kind they are worked intermittently and on a small scale.   Ten miles northeast of Pierce on French Creek, tributary to the North Fork of the Clearwater, are two quartz veins half a mile apart, and called the Golden West and Klondike.   Among the important veins north of Pierce is mentioned the Sweet mine, which from 1894 to 1896 is reported to have produced $40,000.   The Mascot vein is situated 15 miles northwest of Pierce, and was operated during 1899.   The equipment consists of two steam stamps. The veins in the vicinity of Pierce are reported to have a southwesterly dip of about 45°.   Below the shallow zone of oxidation the ore is apt to turn base, as indeed is suggested by the low grade of the placer gold.

Prof. I. C. Russell,[a] who visited the Mascot Hills to the west of Pierce, says that the rocks there "contain numerous small gash veins, but do not reveal the presence of any true lodes or fissure veins."

In 1901 the following properties reported production of gold and silver:

Vi-Vi placers.   May Creek placers.
Last Show placers.   Wild Rose quartz.
Woodchuck placers.   Boise quartz.
Monahassett placers.   Fisher quartz.
Boggs Gulch placers.   Venus quartz.
Missouri Rose placers.

For 1902 the following report is available:

The Wild Rose, owned by Dunn and Wilkinson, Pierce. Idaho, has been operating a 3-stamp mill in 1901 and 1902, and made a good clean up.   Developments amount to 460 feet.,   The Santiago group. owned by Wilsey & Co., of Pierce, has operated a 2-stamp mill the entire season.   The Mascot Mining and Milling Company, of Pierce, has operated its 10-stamp mill part of the season.   The Ozark Company, owned by Gaffney and Pond, of Pierce, also has a mill which has been producing during the season.   The Dewey is operated by Ellis & Co., who are making arrangements for the erection of a 5-stamp mill.   The French Creek Mining and Development Company were erecting a new mill and expected to run it during the winter of 1902–3.

The placer property owned by the American Placer Mining Company, on Oro Fino Creek, was in operation the first part of the year, but closed down in July on account of low water.   This company is installing an elevator plant with 7 miles of ditch and flume, and own about 10 miles of placer ground below Pierce.   Two placer-mining companies are preparing to work French Creek by means of bed-rock flumes.   An elevator was being operated during the summer of 1902 on Gold Creek, 20 miles northwest of Pierce.

From these reports it will be seen that a great number of quartz prospects are actively developed near Pierce.   The aggregate production of the Pierce

---

a Water-Sup. and Irr. Paper No. 54, U. S. Geol. Survey, 1901, p. 128.

mining district now probably exceeds that of Elk City. The output of the camp in 1902 is estimated to be $30,000 from placers and about the same amount from quartz veins. These data have been obtained from Dr. D. T. Day, of the division of mining and mineral resources of the United States Geological Survey, through Mr. V. C. Heikes, field assistant.

MUSSELSHELL AND LOLO CREEKS.

Lolo Creek heads near Rocky Ridge between the Lochsa and the North Fork, and after a tortuous course empties into the main Clearwater at Greers Ferry, a few miles above the mouth of Oro Fino Creek. In its upper course it is joined by Musselshell Creek, which heads in the same vicinity. The latter creek has been mined all the way, although no extraordinary richness is reported. Lolo Creek has been washed for gold in places, but ordinarily contains too many bowlders. The gold, like that of Pierce, is fine in size but of low grade, varying in value from $17 to $15 per ounce. A number of quartz veins are reported from the head of these two creeks, and the district may be considered as the southeasterly extension of the gold-bearing area of Pierce and the Mascot Hills. A prospect called the Pioneer was worked in 1899 and is situated on the low ridges between Musselshell and Lolo creeks, 10 miles east of Weippe. The vein strikes N. 70° E. and dips 85° S.

DEPOSITS BETWEEN CLEARWATER AND SNAKE RIVERS.

In the plateau country extending between Clearwater and Snake rivers the Columbia River lava covers nearly the whole area, though bed rock is occasionally exposed in the deep cuts of the canyons or in points protruding through the lava. No placer mines of importance have been worked in this area. The bottom gravels and bars of Salmon River and of Snake River contain gold, however, and these have been worked to some extent. Along Salmon River, especially where the gold is less fine than along the Snake, placer operations have been carried on in different places during the last few years. The aggregate yield is, however, small.

Near Harpster and Mount Idaho.—Many small quartz veins have been located and partly developed along the South Fork of the Clearwater near Harpster. None of them have, however, proved rich. The rock inclosing them consists of black slate, diorite, and amphibolite. Some miles above Harpster, along the same river and only a few miles east of Mount Idaho, a few veins have been discovered and were prospected in 1899. These veins form a silicified and mineralized streak in greenstone-schists almost at the river level and on the west side of the stream. The strike is northeasterly and the dip 60° SE. The main tunnel shows principally an altered greenstone containing much pyrite.

There are occasional bunches of quartz containing pyrite and chalcopyrite and inclosed in greenstone-schists. Good assays and free gold are reported to have been obtained from the Dewey, which was the main prospect worked at the time of visit. The developments chiefly consisted of a 50-foot tunnel.

Since 1899 a considerable amount of work has been done on these prospects and a great deal of ore has been shipped. The prospects near the river have been united as the Dewey Consolidated Mining and Smelting Company, and up to the present time the development work (tunnels, shafts, raises, and winzes) aggregates 3,000 feet. Nearly 200 tons of rich ore were shipped to smelters during 1900, yielding net $13,000, according to treasurer's report. Mr. Otto Abeling has made a report on the property in 1900 from which it appears that there are four parallel veins, striking N. 28° E. and dipping 75°. On the two outside veins, the Waverly and the West veins, but little work has been done, while the principal developments have been effected on the Dewey and the St. Patrick. The width of the veins is from 23 to 40 feet. The whole distance between the east wall of the Dewey and the west wall of St. Patrick is said to be 180 feet, within which distance all assays show some value. As stated above, the rock is a roughly schistose greenstone and the veins are about parallel to the schistosity. The ore consists of silicified and pyritic greenstone containing streaks and veinlets of quartz. The immediate outcrops of this rock were rather poor, showing little copper and but few dollars in gold. Tunnels driven on both veins soon struck a number of streaks of shipping ore very rich in gold and greater bodies of lower-grade material, both classes containing chalcopyrite in varying but not large amounts. Very little of the gold is free. On the Dewey vein more copper ores were found than on the St. Patrick. A small shoot of massive pyritic ore 3 feet wide yielded 3.4 per cent copper and $5 in gold. In another place a sample across 4 feet yielded 3.1 per cent copper, 12.7 ounces silver, and 4.38 ounces gold, according to Mr. Abeling, who also estimates the ore in sight as 34,650 tons with a value of $361,200.

Higher up on the hill, west of the Dewey, is the similar Evergreen property from which considerable shipments of smelting ore were reported in 1901.

In general character these deposits differ from the ordinary type of gold quartz veins common in this part of the country. They show a certain similarity to the Bully Hill deposits of Shasta County, Cal., and in both cases a certain amount of secondary concentration has probably been active.

Cottonwood Buttes.—The area of old rocks which protrude as low hills above the general surface of the Columbia River lava, a few miles north of Cottonwood on the Camas Prairie, contain several veins of quartz which have been prospected on a small scale. The veins strike north and south and are nearly vertical; the width is reported to be 4 feet, and the valuable metals contained are stated to be silver, gold, and copper. A shaft 115 feet deep has been sunk on the Eckert and Smith vein. As to the actual value of the veins very little is known.

The lower Clearwater.—Few deposits of importance have been discovered on the lower Clearwater River. The only exception, as far as I am aware, consists in a few veins reported to contain gold which have been discovered on Bedrock Creek, approximately 15 miles below Oro Fino and 25 miles above Lewiston. Quartz veins are reported to occur here in calcareous quartzite cropping below the level of the lava plateau three-fourths of a mile above the mouth of the creek.

Canyons of Snake and Salmon rivers.—In the deep trenches of these two rivers south of Lewiston the bed-rock formations below the lava, when exposed, generally consist of highly disturbed and sedimentary rocks accompanied by greenstones of various kind. Neither of the canyons were visited during the present examination. On Salmon River prospects are reported to occur in these rocks near White Bird.

Along Snake River several veins containing copper in promising quantity have lately been discovered near the mouth of the Salmon Canyon. From reports they are often connected with greenstones of dioritic character. A vein known as the Mary K is reported 5 miles below the mouth of Salmon River, near the point where Cottonwood Creek empties into Snake River. At the junction of the Imnaha River, entering the Snake a few miles above the mouth of Salmon River, important discoveries are reported in the rocks which here underlie the thick flows of the Columbia River lava. In this bed-rock series, which extends for several miles up above the junction on both rivers, occur a series of deposits which, from several concurrent reports, must be considered as fissure veins; they contain a rather peculiar combination of chalcopyrite and massive steel-gray hematite. The amount of ore is reported as considerable, although the grade is stated to be low. If this hematite is a primary mineral, as would appear probable from the specimens shown me, it is not likely that the composition of the ore will undergo great change in depth. The veins are reported to extend over from the Imnaha to the Snake, having a general strike of northeast and southwest. Besides copper, the ore contains a small amount of gold and silver.

There are several gravel benches in the lower course of the Salmon above Freedom which have been mined. A short distance below Freedom operations on the low bars were contemplated at the time of my visit in 1897. At the mouth of John Day Creek hydraulic mining has been done on a bar 100 feet above the river, the bed rock being probably 50 feet above the same level. Three-fourths of a mile above the mouth of John Day Creek on the east side of the river is a high bar, the slate bed rock being 80 feet above river level and the top of the sandy gravel 300 feet above the same. This bar was hydraulicked in 1897 by the Victor Mining Company, of Spokane. The pay is concentrated in a thin layer of cemented gravel on the bed rock.

A big gravel bar 200 feet high is situated higher up on the west side opposite Carver's ranch. The limestone bed rock is only a few feet above the river and on it rests cemented gravel, containing very coarse gold of local origin. This deposit has been mined by drifting along the bed rock. Another large bar with much sand lies at a bend of the river a little north of Fiddle Creek. Still another high bar on which some placer mining has been done is situated on the south side of the river a few miles above the great bend at Salmon Point. Above this point there are only smaller low bars but the sands of these contain fine gold everywhere.

COEUR D'ALENE MOUNTAINS.

What little is known about this subject, really outside of the limits of this reconnaissance, is given below. For notes on the geology see page 81.

GOLD DEPOSITS.

The Coeur d'Alene Mountains are not rich in known gold deposits. Placer mines of moderate richness have been worked in the Moose Creek diggings on a branch of the North Fork of the Clearwater, a short distance west of the Montana-Idaho line. Minor deposits have also been worked at the head of the Middle Fork of St. Josephs River and on the Montana side near the head of Cedar Creek halfway between Mullan Pass and Missoula.

The most important deposits, placers as well as quartz veins, are situated at Murray and Delta on the North Fork of the Coeur d'Alene River, 15 miles north of the lead-mining district.

LEAD-SILVER VEINS.

To this class belong the celebrated deposits near Wardner, Wallace, and Mullan on the Coeur d'Alene River,[a] which are estimated to have produced about $60,000,000 of lead and silver since their discovery fifteen years ago. The lowest profitable percentage of lead in the mined ore is said to be about 6 per cent. The ores are subjected to careful concentration in large dressing works.

The deposits are clearly defined fissure veins cutting fine-grained greenish quartzites and quartzitic slates of doubtful (though probably pre-Cambrian) age. The principal ores are galena and zincblende, but there is also much finely distributed pyrite. Chalcopyrite is ordinarily absent. These are practically the only metallic minerals, and recur in all the deposits. The principal gangue mineral is

---

[a] For a brief description of these mines, including technical features, see J. R. Finlay, Trans. Am. Inst. Min. Eng., February and May, 1902.

siderite, accompanied by minor quantities of quartz and barite. Fluorite is absent.
The fissures along which the ore bodies appear are well defined, and sometimes
continuous for one or more miles. The ore bodies do not show much clearly defined
crustification or other evidence of having been deposited in open space. The
siderite appears always as an undoubted product of replacement, while many
veinlets of quartz have in part resulted from the filling of open small fissures.
Evidences of gradual transitions from ore to country rock are abundant, and are
especially prominent in the mines carrying low-grade ore, as, for instance, in the
Helena and Frisco. In the exposures underground, as well as in the specimens and
thin sections, the evidence of replacement is complete and positive.

The greenish-gray fine-grained quartzite which constitutes the prevailing
country rock contains no sulphides when fresh. It is composed of small, rounded,
or subangular quartz grains, closely packed—often, indeed, jointing closely, as in a
normal quartzite. Usually, however, a little sericite, in bunches of small fibers, is
present as cementing material between the grains. There are few other minerals,
exept a little feldspar in clastic grains, small prisms of tourmaline, and some grains
of calcite. Near the veins minute specks of siderite, zincblende, pyrite, and galena
appear in this quartzite; and these scattered grains gradually merge into bodies of
merchantable ore. The thin sections show how the rock near the veins is filled with
small grains of branching and irregular form, which consist of siderite, developed
by attack first upon the groundmass and then upon the grains of clastic quartz.
Accompanying the siderite are small grains of zincblende, cubes of pyrite and
irregularly wiry masses of galena. All these sulphides appear not only in or near
the siderite, but also in the cementing sericite, and in the apparently perfectly fresh
quartz grains.

At a more advanced stage these areas of siderite extend until they join, and
thus completely replace the rock. In the resulting ore lie scattered many small
quartz grains, representing remnants of the clastic constituents of the quartzite.
Occasionally larger masses of zincblende appear to form directly in the quartzite by
metasomatic replacement of the quartz. The sericite in the quartz then disappears,
though once in a while small foils of it may be detected. During the transition
stage, seams and narrow veinlets in the altering rock are filled with sericite, apparently segregated there, when driven out from the main mass. In other specimens
from the Helena and Frisco mine, the replacing siderite has a strong tendency to
idiomorphic development. Imperfect rhombohedral forms are often seen, sometimes
cutting straight across the clastic grains. Certain specimens from the Bunker Hill
and Sullivan mine show quartzose grayish masses of irregular outline, and apparently
merging gradually into the normal greenish quartzite. These quartzose masses
consist of very irregular interlocking grains of quartz, not in the least similar to the

quartz usually deposited by processes of filling, but having every appearance of resulting from the silicification of the quartzite. This silicified portion contains irregular grains of pyrite, galena, and brown zincblende, with a very little siderite.

The process, as outlined, is remarkable, as involving a metasomatic replacement of quartz by siderite, pyrite, galena, and zincblende, and is the only clearly defined occurrence of this kind of which I am aware.

Coming from the west the first deposits are found near the town of Wardner. The Bunker Hill lode is here the principal source of production of the Wardner district, estimated by Mr. J. R. Finlay to be $17,500,000. This remarkable vein system is continuous for 2 or 3 miles; its principal feature is a well-defined fault plane contained in quartzite and striking N. 42° W., the dip being 38° SW. The ore forms irregular masses above this fault plane or "foot wall." The mines are worked by tunnels, a deepest level being recently reached by a long 8 by 9 foot tunnel from near the Coeur d'Alene River, striking the veins 700 feet below the previously lowest tunnel level on the Bunker Hill and Sullivan mines. The croppings along the foot wall is indicated by a heavy iron cap stoped 50 feet wide at the surface. The ore is mainly galena, with a little pyrite and blende; the concentrates carry about 25 ounces of silver. There is practically no arsenic, antimony, or gold. Some galena appeared in the croppings, but also much cerussite in beautiful bunches and this secondary mineral continues in places as far as 200 feet below the surface.

Although the foot wall is so clearly defined by a fault plane, accompanied by a few inches of clayey gouge, the hanging wall of the ore bodies is extremely ill defined and in many places the mineralization simply fades out gradually. Some of the ore shoots lie on the foot wall, but others reach out as much as 200 feet horizontally from this plane. Some of the masses of galena, instead of strictly following the course of the foot wall, may locally follow a stratification plane and then jump back in offsets so as to preserve the general parallelism with the wall. The thickness of the ore bodies is as great as 90 feet; within the pay shoots the galena occurs irregularly; there are some massive bodies as much as 10 feet thick, which may change their outlines rapidly and irregularly and which are sometimes separated by horizontal joint planes from the barren quartzite. Most commonly, in the poorer ores, the galena occurs as small and very irregular seams associated with some quartz and siderite. The greatest ore body in the Bunker Hill and Sullivan vein had a maximum thickness of 90 feet, was several hundred feet long, and extended across a vertical distance of 170 feet. Its ore averaged 25 per cent lead. The ore stopped against a local hanging wall which lower down joined the principal foot wall.

Near Gem and Burke, 10 miles farther up the river and near a granite boss which here is intruded into the steeply dipping quartzites and clay slates, is another important vein system, generally called the "Canyon Creek lodes." The production of this system, which include such important mines as the Helena-Frisco, Tiger-Poorman, Mammoth Standard, Morning, You Like, and others, is estimated as $42,500,000 (J. R. Finlay). One system of vein strikes from Nine-mile Creek across to Gem in an east-southeast direction, and thence across toward Mullan. Two parallel systems are found 2 and 4 miles north of Gem. The veins of the "Canyon Creek lodes" are simpler in structure than the Wardner deposits, but like those are typical replacement veins along shear zones of greater or smaller width.

The Helena-Frisco, a representative mine, produces about 600 tons of ore per day, which is concentrated in proportions of 9 to 1, yielding 50 to 70 tons of concentrates. The mine is opened by a shaft 1,800 feet deep. The country rock is a fine-grained greenish-gray quartzite. The vein which cuts the quartzite in strike and dip has a course of N. 80° W., and a variable southwest dip averaging 70°, and gradually straightening on lower levels. The ore averages 10 to 12 feet in width, and is taken out uniformly. The foot wall is very poorly defined; the hanging wall is also often indistinct, though usually roughly indicated. Sometimes every indication of geometrical walls is absent. The rock is hard and compact, even within the walls. Within the width of 12 feet are irregular streaks of fine-grained steely galena and blende, occasionally a couple of feet thick. Pyrite, and in places chalcopyrite, are sparingly represented. The gangue is finely divided siderite and a little quartz, the latter in narrow and irregular seams. The vein is clearly the result of replacement of quartzite by galena, zinc blende, and siderite, along tight fault planes. The vein system of which the Helena-Frisco forms a part continues for 6 miles and is faulted at several places by later fissures striking north-northeast.

The Morning and You Like are situated on the eastern end of this vein system, high up on the hill above Mullan. A tunnel 9,000 feet long and 900 feet below the working tunnel used in 1899 is projected to tap the vein from the mill level. The ore shoots, which are said to be strong, are 600 to 700 feet in length and 10 to 40 feet in width. The present capacity of this group is 700 tons of ore per twenty-four hours.

LIGNITE.

Coal of a lignitic character has been found in two widely distant parts of the area under examination. In neither case are the developments extensive enough to assure their commercial importance. At the head of the South Fork of Bitterroot River lignite has been found on Coal Creek, a small tributary from the west which enters the South Fork 10 miles above the mouth of the West or Nez Perce Fork. The deposit is connected with rhyolite, and appears, in fact, interbedded together with clays and sands in the upper part of the rhyolitic flows. Mr. J. B. Leiberg[a] states that the deposit occupies an area of about 500 acres, and that two seams have been discovered, the upper one with a thickness of 12 feet, while the extent of the lower is unkown. Coal Creek was visited by Mr. G. W. Stose, who reports that the rhyolite exposed in the bottom of the canyon continues practically all the way up to an elevation of 1,000 feet above the main stream. Two tunnels, 50 feet apart and at slightly different levels, the upper one boarded up, are the only developments. The coal seam was only partly exposed; 2½ feet of coal were seen, with a 3-inch seam of clay in the middle. The dip of the strata seems to be 22°. The roof of the coal consists of a clay full of vegetable matter, such as fern-like leaves, grasses, etc. Above this clay follows a hard and compact rhyolite tuff. The extent of the coal lands is problematic, and I am unable to say whether Mr. Leiberg's estimate is correct. At any rate, it is clear that the extent of the territory underlain by coal is small, and the well-known uncertainty of thickness and quality in other fields similar to this make predictions of doubtful value. A sample of the coal was analyzed by Dr. W. F. Hillebrand, with the following result:

Analysis of coal from near head of South Fork of the Bitterroot River.

	I.	II.
Moisture at 105°	7.21	7.21
Volatile combustibles	56.09	41.84
Fixed carbon	27.09	37.61
Ash	9.61	13.34
	100.00	100.00

The first column represents the composition as usually determined, the ash being found in the same portion as the volatile combustibles and fixed carbon. But the coal sparks enormously, and much ash and fixed carbon are ejected with the volatile combustibles.

The second column gives the true ash contents and the corrected volatile combustibles and fixed carbon, calculated on the assumption that the ratio of fixed carbon to ash in the ejected matter is the same as that in the coke of I. The coke is thoroughly sandy. The ash is pale buff colored.

The coal is black in color, and crumbles easily on exposure to the air. The analysis shows it to be a lignite of very fair quality, approaching a bituminous coal in composition. From its association with the rhyolite, the conclusion is strong that it is probably of Miocene age. A little lignite is reported to have

[a] Nineteenth Ann. Rept. U. S. Geol. Survey, pt. 5, 1898, p. 265.

been found in the upper area of rhyolite, 10 miles higher up on the South Fork, where a similar succession of rhyolitic flows and tuffs is exposed.

The only other known occurrence is that on the west side of the Clearwater Mountains in the canyon of the Clearwater River. This coal, which occurs in a series of sedimentary beds interstratified with the Columbia River lava, has been described by Prof. I. C. Russell.[a] Professor Russell states that three seams of lignite from 2 to 20 inches in thickness and separated by a few feet of sandstone have been found half mile below Oro Fino on the south side of Clearwater River and at an elevation of about 200 feet above the same. The exposed beds are believed to be part of a landslide, and the true position of the beds is suggested to be 500 feet higher. Lignite is reported to occur at several localities on Oro Fino Creek (all of it probably belonging to approximately the same bed) within a distance of several miles in the middle portion of its course. In sec. 12, T. 36 N., R. 4 W. an exposure of this kind was examined by Professor Russell, who found 20 inches of lignite embedded in sandstone and clay. In another place near this exposure a shaft has exposed 27 inches of lignite. Here again landslides are believed to have occurred, and the true position should be looked for at an elevation of between 400 and 600 feet above the creek. The lignite-bearing rocks no doubt underlie a wide extent of the adjacent plateaus.

Similar exposures are also reported from several places in Little Canyon to the south of Oro Fino and on the west side of the river. At the place which was examined by Professor Russell no lignite but only carbonaceous shales were observed.

MINERAL SPRINGS.

A limited number of generally hot mineral springs are scattered over the area examined without special regularity and without showing any connection with lines of structure or volcanic flows. Beginning on the north, hot springs are found at the head of Lolo Fork and about 6 miles north of Lolo Pass. These springs, which yield a weak mineral water of high temperature and are of sufficient volume to supply a bathing establishment, break through granite not far from the contact with the sedimentary series. Going over to the Clearwater side, hot springs of similar nature are found at Jerry Johnson's on the Lochsa Fork at an elevation of 3,300 feet. They issue at different places a short distance below Warm Springs Creek and on the north side of the river. No hot springs are known to exist along the foot of the Bitterroot Range, but in the eastern foothills, about 12 miles southeast of Hamilton, hot waters issue at the place known as Weeping Child Hot Springs.

No hot springs are known on the South Fork of the Bitterroot River, but a cold soda spring occurs on a branch of the West or Nez Perce Fork of the Bitterroot

---

[a] Water-Sup. and Irr. Paper No. 54. U. S. Geol. Survey, 1901, p. 122.

River 6 or 7 miles above its mouth.   The rock is here a pegmatitic granite.   In the Salmon River drainage hot springs issue on Horse Creek at an elevation of about 6,000 feet.   The water, which has a temperature of 110° F., has no taste or odor. Another and somewhat hotter spring issues half a mile below the upper place. Farther west in the Clearwater drainage a hot spring is stated to occur on the headwaters of the Red River, a branch of the South Fork.

It is a somewhat peculiar fact that no hot waters are known from the large area occupied by the Columbia River lava between the Clearwater and the Snake River.

SUMMARY.

The region described is bordered on the south by the Salmon River and on the north by the North Fork of the Clearwater.   The fertile Bitterroot Valley is at the eastern foot of the imposing Bitterroot Range, which attains an elevation of 11,000 feet, and westward merges into the great dissected high plateau of the Clearwater Mountains, which at their western edge descend rather abruptly to the lower plateaus of Camas and Cold Spring prairies, forming part of the great Columbia River lava plateau.   This plateau has a general elevation of 2,500 to 3,000 feet, and is built up of horizontal lava flows.   From great glacial cirques in the western slopes of the Bitterroot Range the Salmon River and the several forks of the Clearwater River find their way westward in canyons from 3,000 to 5,000 feet deep. The canyon of the Salmon especially is remarkable for its great length and depth. In the lower plateau country these rivers flow in more sharply incised but less deep canyons, which continue to their junction with the master stream, the Snake River. The area is a wild and very sparsely populated mountain region, heavily timbered except on the highest ridges, which usually show clear evidence of glacial action.

The geology is comparatively simple.   The main Bitterroot Range and the larger part of the Clearwater Mountains consist of a granite, or, defining it more correctly, of a quartz-monzonite mass—the northward continuation of the great batholith of the same rock which occupies so large an area in south-central Idaho. In the latter region this intrusive mass is of post-Carboniferous and probably late Mesozoic age, and there is no reason to believe that the granite of the Clearwater and the Bitterroot is of different age.

Along the whole eastern slope of the Bitterroot Mountains this granite is made schistose by pressure and forms a zone a few miles in width and 60 miles long, following the front of the range.   This schistose zone is probably due to a great fault which, like the schistosity, dips about 18° E.   Otherwise the granite is generally massive and but little altered.   Several smaller areas of a much older gneiss (pre-Cambrian?) occur in the Clearwater Mountains, the largest appearing near Elk City.   The granite is intrusive in this gneiss.   Along Lolo Fork, at the northern end of the Bitterroot Mountains, and near the head of the Bitterroot River are areas of probably Cambrian or pre-Cambrian quartzites and slates, into which the granite is also intrusive. Finally, along the western foot of the Clearwater Mountains, near Harpster and Mount Idaho, occur slates, limestones, and greenstones, which clearly form the northeasterly extension of the similar sedimentary series of the Seven Devils and the lower Salmon River. This series is believed to be of early Mesozoic age and the granite of the Clearwater Mountains is intrusive in it.

The geological history of this region probably begins with the deposition of pre-Cambrian sediments on a basement of gneiss. Much later, in early Mesozoic time, sediments, accompanied by basic lavas, were laid down on the western side of the area described. Granitic intrusions followed or accompanied a great uplift of these sediments, and an active erosion reduced the surface to moderate relief. All this is believed to have been accomplished before the beginning of the Cenozoic era. A second uplift raised this reduced topography to the present level of the great Clearwater Plateau, which probably extended far into Montana. On the west this plateau is probably separated from the Columbia River Valley by a dislocation, while the Bitterroot Range was differentiated by the remarkable flat fault, 60 miles long, which now forms its eastern slope. The deep canyons of the Clearwater and the Salmon were excavated before the eruption of the Columbia River basalt. The movement along the great Bitterroot fault plane has apparently continued until a recent date. The vertical movement along the Bitterroot fault is from 4,000 to 6,000 feet, and the horizontal component at least 2 miles in length. The fault appears to be a normal one. A twofold movement has probably taken place, by which the foot wall has been raised and at the same time the hanging wall depressed. The Bitterroot Valley represents the result of the downthrow, the Bitterroot Range the result of the upward movement. This dislocation, remarkable in many ways, probably is the most important structural feature discussed in this report.

During Miocene times, finally, followed the great outflow of basaltic lavas at the foot of the Clearwater Mountains. These swiftly succeeding flows formed the plateau of Columbia River lava, their outbursts damming the old canyons and causing many changes of river courses, especially along the lower Salmon and Clearwater rivers. The whole course of the Snake River below Weiser, Idaho, has been changed by their eruption. Slow movements have taken place since the Miocene, during which time the main body of the Columbia River basalts were erupted. Besides local warping, such as shown in the Craig Mountains and the Uniontown Plateau, near Lewiston, Idaho, it is believed that a general subsidence of a large part of the lava-covered area has taken place. Other parts, however, as, for instance, the high plateau near the Seven Devils, through which Snake River has cut its deep canyons, have probably been raised, the canyon cutting keeping pace with the uplift.

The latest event of geological interest is the glaciation. During early Quaternary times a continuous ice sheet covered the whole of the Bitterroot

Range for a distance of 60 miles from north to south. Isolated high parts of the Clearwater Mountains were likewise centers of local glaciation. The lower limit of the glaciated areas is an elevation of 7,000 feet, though ice tongues on both sides of the mountains reached down to elevation of 4,000 feet.

The valuable minerals contained in this area are chiefly confined to the western slope of the Clearwater Mountains. The gold deposits are the most important, and occur as fissure veins and gravels. At the foot of the Bitterroot Mountains, about 12 miles north of Hamilton, Mont., is the remarkable Curlew mine, where a body of galena was found on a fissure having limestone and quartzite as its foot wall and the gravels of the Bitterroot Valley as hanging wall. On the upper South Fork of Bitterroot River prospects of copper, silver, and silver-lead ores have been worked. Hughes Creek, in the same vicinity, contains gold-bearing gravels which have been worked for many years.

Elk City, founded in 1862, is one of the centers of the gold-mining industry on the west slope. Gold to the value of several million dollars was extracted here from late Tertiary bench gravels, and placer mining is still prosecuted on a small scale. Quartz veins have also been recently opened in this old mining district. The well-defined quartz veins of Buffalo Hump, discovered in 1897, are actively worked, though not as yet on a very large scale. These deposits occur in granitic rocks, accompanied by contact metamorphic slates.

Florence, located 40 miles southwest of Elk City, is another of the old placer camps, discovered about 1861. The gravels were very rich, and the camp has yielded gold to the value of many million dollars. At present a few placers and quartz mines are worked there also. The bed rock is granite.

Pierce, on Oro Fino Creek, about 50 miles north-northwest of Elk City, also a well-known placer camp, was founded in 1861, and its gravels have been continuously worked since then. The present yield from these placers is probably about $30,000 per annum. The gravels occur as creek alluvium and as benches up to 500 feet above the present drainage level. Quartz veins have also been found near Pierce, and many of them are being worked on a small scale. The veins occur in gneisses and schists.

The total production of this gold-bearing region, extending at the western foot of the Clearwater Mountains, from the North Fork of the Clearwater to Salmon River, is at the present time probably about $160,000 per annum.

Coal of fair quality has been found on the upper South Fork of the Bitterroot Valley, and the beds are here associated with rhyolite. The developments have not as yet shown the extent of this occurrence.

Lignite of fair quality also occurs on Oro Fino Creek below Pierce, but the beds are probably not thick enough to allow profitable working.

INDEX.

O

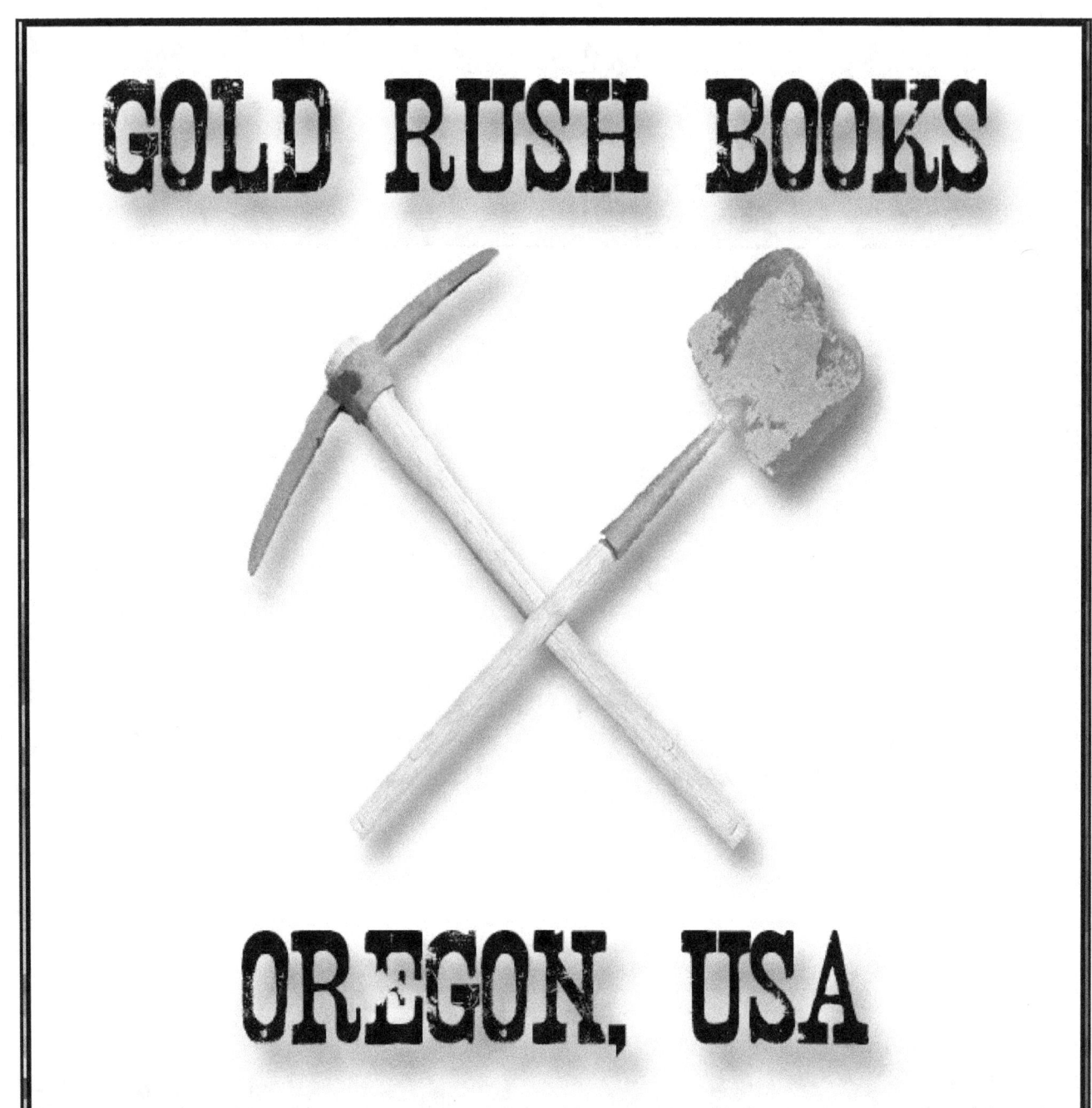

GOLD RUSH BOOKS

OREGON, USA

www.GoldMiningBooks.com

Books On Mining

Visit: www.goldminingbooks.com to order your copies or ask your favorite book seller to offer them.

Mining Books by Kerby Jackson

Gold Dust: Stories From Oregon's Mining Years - Oregon mining historian and prospector, Kerby Jackson, brings you a treasure trove of seventeen stories on Southern Oregon's rich history of gold prospecting, the prospectors and their discoveries, and the breathtaking areas they settled in and made homes. **5" X 8", 98 ppgs. Retail Price: $11.99**

The Golden Trail: More Stories From Oregon's Mining Years - In his follow-up to "Gold Dust: Stories of Oregon's Mining Years", this time around, Jackson brings us twelve tales from Oregon's Gold Rush, including the story about the first gold strike on Canyon Creek in Grant County, about the old timers who found gold by the pail full at the Victor Mine near Galice, how Iradel Bray discovered a rich ledge of gold on the Coquille River during the height of the Rogue River War, a tale of two elderly miners on the hunt for a lost mine in the Cascade Mountains, details about the discovery of the famous Armstrong Nugget and others. **5" X 8", 70 ppgs. Retail Price: $10.99**

Oregon Mining Books

Geology and Mineral Resources of Josephine County, Oregon - Unavailable since the 1970's, this important publication was originally compiled by the Oregon Department of Geology and Mineral Industries and includes important details on the economic geology and mineral resources of this important mining area in South Western Oregon. Included are notes on the history, geology and development of important mines, as well as insights into the mining of gold, copper, nickel, limestone, chromium and other minerals found in large quantities in Josephine County, Oregon. **8.5" X 11", 54 ppgs. Retail Price: $9.99**

Mines and Prospects of the Mount Reuben Mining District - Unavailable since 1947, this important publication was originally compiled by geologist Elton Youngberg of the Oregon Department of Geology and Mineral Industries and includes detailed descriptions, histories and the geology of the Mount Reuben Mining District in Josephine County, Oregon. Included are notes on the history, geology, development and assay statistics, as well as underground maps of all the major mines and prospects in the vicinity of this much neglected mining district. **8.5" X 11", 48 ppgs. Retail Price: $9.99**

The Granite Mining District - Notes on the history, geology and development of important mines in the well known Granite Mining District which is located in Grant County, Oregon. Some of the mines discussed include the Ajax, Blue Ribbon, Buffalo, Continental, Cougar-Independence, Magnolia, New York, Standard and the Tillicum. Also included are many rare maps pertaining to the mines in the area. **8.5" X 11", 48 ppgs. Retail Price: $9.99**

Ore Deposits of the Takilma and Waldo Mining Districts of Josephine County, Oregon - The Waldo and Takilma mining districts are most notable for the fact that the earliest large scale mining of placer gold and copper in Oregon took place in these two areas. Included are details about some of the earliest large gold mines in the state such as the Llano de Oro, High Gravel, Cameron, Platerica, Deep Gravel and others, as well as copper mines such as the famous Queen of Bronze mine, the Waldo, Lily and Cowboy mines. This volume also includes six maps and 20 original illustrations. **8.5" X 11", 74 ppgs. Retail Price: $9.99**

Metal Mines of Douglas, Coos and Curry Counties, Oregon - Oregon mining historian Kerby Jackson introduces us to a classic work on Oregon's mining history in this important re-issue of Bulletin 14C Volume 1, otherwise known as the Douglas, Coos & Curry Counties, Oregon Metal Mines Handbook. Unavailable since 1940, this important publication was originally compiled by the Oregon Department of Geology and Mineral Industries includes detailed descriptions, histories and the geology of over 250 metallic mineral mines and prospects in this rugged area of South West Oregon. **8.5" X 11", 158 ppgs. Retail Price: $19.99**

Metal Mines of Jackson County, Oregon - Unavailable since 1943, this important publication was originally compiled by the Oregon Department of Geology and Mineral Industries includes detailed descriptions, histories and the geology of over 450 metallic mineral mines and prospects in Jackson County, Oregon. Included are such famous gold mining areas as Gold Hill, Jacksonville, Sterling and the Upper Applegate. **8.5" X 11", 220 ppgs. Retail Price: $24.99**

Metal Mines of Josephine County, Oregon - Oregon mining historian Kerby Jackson introduces us to a classic work on Oregon's mining history in this important re-issue of Bulletin 14C, otherwise known as the Josephine County, Oregon Metal Mines Handbook. Unavailable since 1952, this important publication was originally compiled by the Oregon Department of Geology and Mineral Industries includes detailed descriptions, histories and the geology of over 500 metallic mineral mines and prospects in Josephine County, Oregon. **8.5" X 11", 250 ppgs. Retail Price: $24.99**

Metal Mines of North East Oregon - Oregon mining historian Kerby Jackson introduces us to a classic work on Oregon's mining history in this important re-issue of Bulletin 14A and 14B, otherwise known as the North East Oregon Metal Mines Handbook. Unavailable since 1941, this important publication was originally compiled by the Oregon Department of Geology and Mineral Industries and includes detailed descriptions, histories and the geology of over 750 metallic mineral mines and prospects in North Eastern Oregon. **8.5" X 11", 310 ppgs. Retail Price: $29.99**

Metal Mines of North West Oregon - Oregon mining historian Kerby Jackson introduces us to a classic work on Oregon's mining history in this important re-issue of Bulletin 14D, otherwise known as the North West Oregon Metal Mines Handbook. Unavailable since 1951, this important publication was originally compiled by the Oregon Department of Geology and Mineral Industries and includes detailed descriptions, histories and the geology of over 250 metallic mineral mines and prospects in North Western Oregon. **8.5" X 11", 182 ppgs. Retail Price: $19.99**

Mines and Prospects of Oregon - Mining historian Kerby Jackson introduces us to a classic mining work by the Oregon Bureau of Mines in this important re-issue of The Handbook of Mines and Prospects of Oregon. Unavailable since 1916, this publication includes important insights into hundreds of gold, silver, copper, coal, limestone and other mines that operated in the State of Oregon around the turn of the 19th Century. Included are not only geological details on early mines throughout Oregon, but also insights into their history, production, locations and in some cases, also included are rare maps of their underground workings. **8.5" X 11", 314 ppgs. Retail Price: $24.99**

Lode Gold of the Klamath Mountains of Northern California and South West Oregon
(See California Mining Books)

Mineral Resources of South West Oregon - Unavailable since 1914, this publication includes important insights into dozens of mines that once operated in South West Oregon, including the famous gold fields of Josephine and Jackson Counties, as well as the Coal Mines of Coos County. Included are not only geological details on early mines throughout South West Oregon, but also insights into their history, production and locations. **8.5" X 11", 154 ppgs. Retail Price: $11.99**

Chromite Mining in The Klamath Mountains of California and Oregon
(See California Mining Books)

Southern Oregon Mineral Wealth - Unavailable since 1904, this rare publication provides a unique snapshot into the mines that were operating in the area at the time. Included are not only geological details on early mines throughout South West Oregon, but also insights into their history, production and locations. Some of the mining areas include Grave Creek, Greenback, Wolf Creek, Jump Off Joe Creek, Granite Hill, Galice, Mount Reuben, Gold Hill, Galls Creek, Kane Creek, Sardine Creek, Birdseye Creek, Evans Creek, Foots Creek, Jacksonville, Ashland, the Applegate River, Waldo, Kerby and the Illinois River, Althouse and Sucker Creek, as well as insights into local copper mining and other topics. **8.5" X 11", 64 ppgs. Retail Price: $8.99**

Geology and Ore Deposits of the Takilma and Waldo Mining Districts - Unavailable since the 1933, this publication was originally compiled by the United States Geological Survey and includes details on gold and copper mining in the Takilma and Waldo Districts of Josephine County, Oregon. The Waldo and Takilma mining districts are most notable for the fact that the earliest large scale mining of placer gold and copper in Oregon took place in these two areas. Included in this report are details about some of the earliest large gold mines in the state such as the Llano de Oro, High Gravel, Cameron, Platerica, Deep Gravel and others, as well as copper mines such as the famous Queen of Bronze mine, the Waldo, Lily and Cowboy mines. In addition to geological examinations, insights are also provided into the production, day to day operations and early histories of these mines, as well as calculations of known mineral reserves in the area. This volume also includes six maps and 20 original illustrations. **8.5" X 11", 74 ppgs. Retail Price: $9.99**

Gold Mines of Oregon - Oregon mining historian Kerby Jackson introduces us to a classic work on Oregon's mining history in this important re-issue of Bulletin 61, otherwise known as "Gold and Silver In Oregon". Unavailable since 1968, this important publication was originally compiled by geologists Howard C. Brooks and Len Ramp of the Oregon Department of Geology and Mineral Industries and includes detailed descriptions, histories and the geology of over 450 gold mines Oregon. Included are notes on the history, geology and gold production statistics of all the major mining areas in Oregon including the Klamath Mountains, the Blue Mountains and the North Cascades. While gold is where you find it, as every miner knows, the path to success is to prospect for gold where it was previously found. **8.5" X 11", 344 ppgs. Retail Price: $24.99**

Mines and Mineral Resources of Curry County Oregon - Originally published in 1916, this important publication on Oregon Mining has not been available for nearly a century. Included are rare insights into the history, production and locations of dozens of gold mines in Curry County, Oregon, as well as detailed information on important Oregon mining districts in that area such as those at Agness, Bald Face Creek, Mule Creek, Boulder Creek, China Diggings, Collier Creek, Elk River, Gold Beach, Rock Creek, Sixes River and elsewhere. Particular attention is especially paid to the famous beach gold deposits of this portion of the Oregon Coast. **8.5" X 11", 140 ppgs. Retail Price: $11.99**

Chromite Mining in South West Oregon - Originally published in 1961, this important publication on Oregon Mining has not been available for nearly a century. Included are rare insights into the history, production and locations of nearly 300 chromite mines in South Western Oregon. **8.5" X 11", 184 ppgs. Retail Price: $14.99**

Mineral Resources of Douglas County Oregon - Originally published in 1972, this important publication on Oregon Mining has not been available for nearly forty years. Included are rare insights into the geology, history, production and locations of numerous gold mines and other mining properties in Douglas County, Oregon. **8.5" X 11", 124 ppgs. Retail Price: $11.99**

Mineral Resources of Coos County Oregon - Originally published in 1972, this important publication on Oregon Mining has not been available for nearly forty years. Included are rare insights into the geology, history, production and locations of numerous gold mines and other mining properties in Coos County, Oregon. **8.5" X 11", 100 ppgs. Retail Price: $11.99**

Mineral Resources of Lane County Oregon - Originally published in 1938, this important publication on Oregon Mining has not been available for nearly seventy five years. Included are extremely rare insights into the geology and mines of Lane County, Oregon, in particular in the Bohemia, Blue River, Oakridge, Black Butte and Winberry Mining Districts. **8.5" X 11", 82 ppgs. Retail Price: $9.99**

Mineral Resources of the Upper Chetco River of Oregon: Including the Kalmiopsis Wilderness - Originally published in 1975, this important publication on Oregon Mining has not been available for nearly forty years. Withdrawn under the 1872 Mining Act since 1984, real insight into the minerals resources and mines of the Upper Chetco River has long been unavailable due to the remoteness of the area. Despite this, the decades of battle between property owners and environmental extremists over the last private mining inholding in the area has continued to pique the interest of those interested in mining and other forms of natural resource use. Gold mining began in the area in the 1850's and has a rich history in this geographic area, even if the facts surrounding it are little known. Included are twenty two rare photographs, as well as insights into the Becca and Morning Mine, the Emmly Mine (also known as Emily Camp), the Frazier Mine, the Golden Dream or Higgins Mine, Hustis Mine, Peck Mine and others. **8.5" X 11", 64 ppgs. Retail Price: $8.99**

Gold Dredging in Oregon - Originally published in 1939, this important publication on Oregon Mining has not been available for nearly seventy five years. Included are extremely rare insights into the history and day to day operations of the dragline and bucketline gold dredges that once worked the placer gold fields of South West and North East Oregon in decades gone by. Also included are details into the areas that were worked by gold dredges in Josephine, Jackson, Baker and Grant counties, as well as the economic factors that impacted this mining method. This volume also offers a unique look into the values of river bottom land in relation to both farming and mining, in how farm lands were mined, re-soiled and reclamated after the dredges worked them. Featured are hard to find maps of the gold dredge fields, as well as rare photographs from a bygone era. **8.5" X 11", 86 ppgs. Retail Price: $8.99**

Quick Silver Mining in Oregon - Originally published in 1963, this important publication on Oregon Mining has not been available for over fifty years. This publication includes details into the history and production of Elemental Mercury or Quicksilver in the State of Oregon. **8.5" X 11", 238 ppgs. Retail Price: $15.99**

Mines of the Greenhorn Mining District of Grant County Oregon - Originally published in 1948, this important publication on Oregon Mining has not been available for over sixty five years. In this publication are rare insights into the mines of the famous Greenhorn Mining District of Grant County, Oregon, especially the famous Morning Mine. Also included are details on the Tempest, Tiger, Bi-Metallic, Windsor, Psyche, Big Johnny, Snow Creek, Banzette and Paramount Mines, as well as prospects in the vicinities in the famous mining areas of Mormon Basin, Vinegar Basin and Desolation Creek. Included are hard to find mine maps and dozens of rare photographs from the bygone era of Grant County's rich mining history. **8.5" X 11", 72 ppgs. Retail Price: $9.99**

Geology of the Wallowa Mountains of Oregon: Part I (Volume 1) - Originally published in 1938, this important publication on Oregon Mining has not been available for nearly seventy five years. Included are details on the geology of this unique portion of North Eastern Oregon. This is the first part of a two book series on the area. Accompanying the text are rare photographs and historic maps. **8.5" X 11", 92 ppgs. Retail Price: $9.99**

Geology of the Wallowa Mountains of Oregon: Part II (Volume 2) - Originally published in 1938, this important publication on Oregon Mining has not been available for nearly seventy five years. Included are details on the geology of this unique portion of North Eastern Oregon. This is the first part of a two book series on the area. Accompanying the text are rare photographs and historic maps. **8.5" X 11", 94 ppgs. Retail Price: $9.99**

Field Identification of Minerals For Oregon Prospectors - Originally published in 1940, this important publication on Oregon Mining has not been available for nearly seventy five years. Included in this volume is an easy system for testing and identifying a wide range of minerals that might be found by prospectors, geologists and rockhounds in the State of Oregon, as well as in other locales. Topics include how to put together your own field testing kit and how to conduct rudimentary tests in the field. This volume is written in a clear and concise way to make it useful even for beginners. **8.5" X 11", 158 ppgs. Retail Price: $14.99**

The Bohemia Mining District of Oregon - Originally published in 1900, this important publication on Oregon Mining has not been available for over a century. Included in this volume are important insights into the famous Bohemia Mining District of Oregon, including the histories and locations of important gold mines in the area such as the Ophir Mine, Clarence, Acturas, Peek-a-boo, White Swan, Combination Mine, the Musick Mine, The California, White Ghost, The Mystery, Wall Street, Vesuvius, Story, Lizzie Bullock, Delta, Elsie Dora, Golden Slipper, Broadway, Champion Mine, Knott, Noonday, Helena, White Wings, Riverside and others. Also included are notes on the nearby Blue River Mining District. **8.5" X 11", 58 ppgs. Retail Price: $9.99**

The Gold Fields of Eastern Oregon - Unavailable since 1900, this publication was originally compiled by the Baker City Chamber of Commerce Offering important insights into the gold mining history of Eastern Oregon, "The Gold Fields of Eastern Oregon" sheds a rare light on many of the gold mines that were operating at the turn of the 19th Century in Baker County and Grant County in North Eastern Oregon. Some of the areas featured include the Cable Cove District, Baisely-Elhorn, Granite, Red Boy, Bonanza, Susanville, Sparta, Virtue, Vaughn, Sumpter, Burnt River, Rye Valley and other mining districts. Included is basic information on not only many gold mines that are well known to those interested in Eastern Oregon mining history, but also many mines and prospects which have been mostly lost to the passage of time. Accompanying are numerous rare photos **8.5" X 11", 78 ppgs. Retail Price: $10.99**

Gold Mining in Eastern Oregon - Originally published in 1938, this important publication on Oregon Mining has not been available for over a century. Included in this volume are important insights into the famous mining districts of Eastern Oregon during the late 1930's. Particular attention is given to those gold mines with milling and concentrating facilities in the Greenhorn, Red Boy, Alamo, Bonanza, Granite, Cable Cove, Cracker Creek, Virtue, Keating, Medical Springs, Sanger, Sparta, Chicken Creek, Mormon Basin, Connor Creek, Cornucopia and the Bull Run Mining Districts. Some of the mines featured include the Ben Harrison, North Pole-Columbia, Highland Maxwell, Baisley-Elkhorn, White Swan, Balm Creek, Twin Baby, Gem of Sparta, New Deal, Gleason, Gifford-Johnson, Cornucopia, Record, Bull Run, Orion and others. Of particular interest are the mill flow sheets and descriptions of milling operations of these mines. **8.5" X 11", 68 ppgs. Retail Price: $8.99**

The Gold Belt of the Blue Mountains of Oregon - Originally published in 1901, this important publication on Oregon Mining has not been available for over a century. Included in this volume are rare insights into the gold deposits of the Blue Mountains of North East Oregon, including the history of their early discovery and early production. Extensive details are offered on this important mining area's mineralogy and economic geology, as well as insights into nearby gold placers, silver deposits and copper deposits. Featured are the Elkhorn and Rock Creek mining districts, the Pocahontas district, Auburn and Minersville districts, Sumpter and Cracker Creek, Cable Cove, the Camp Carson district, Granite, Alamo, Greenhorn, Robinsonville, the Upper Burnt River Valley and Bonanza districts, Susanville, Quartzburg, Canyon Creek, Virtue, the Copper Butte district, the North Powder River, Sparta, Eagle Creek, Cornucopia, Pine Creek, Lower Powder River, the Upper Snake River Canyon, Rye Valley, Lower Burnt River Valley, Mormon Basin, the Malheur and Clarks Creek districts, Sutton Creek and others. Of particular interest are important details on numerous gold mines and prospects in these mining districts, including their locations, histories, geology and other important information, as well as information on silver, copper and fire opal deposits. **8.5" X 11", 250 ppgs. Retail Price: $24.99**

Mining in the Cascades Range of Oregon - Originally published in 1938, this important publication on Oregon Mining has not been available for over seventy five years. Included in this volume are rare insights into the gold mines and other types of metal mines in the Cascades Mountain Range of Oregon. Some of the important mining areas covered include the famous Bohemia Mining District, the North Santiam Mining District, Quartzville Mining District, Blue River Mining District, Fall Creek Mining District, Oakridge District, Zinc District, Buzzard-Al Sarena District, Grand Cove, Climax District and Barron Mining District. Of particular interest are important details on over 100 mines and prospects in these mining districts, including their locations, histories, geology and other important information. 8.5" X 11", 170 ppgs. Retail Price: $14.99

Beach Gold Placers of the Oregon Coast - Originally published in 1934, this important publication on Oregon Mining has not been available for over 80 years. Included in this volume are rare insights into the beach gold deposits of the State of Oregon, including their locations, occurance, composition and geology. Of particular interest is information on placer platinum in Oregon's rich beach deposits. Also included are the locations and other information on some famous Oregon beach mines, including the Pioneer, Eagle, Chickamin, Iowa and beach placer mines north of the mouth of the Rogue River. 8.5" X 11", 60 ppgs. Retail Price: $8.99

Mineralogical Composition of the Sands of the Oregon Coast: From Coos Bay to the Columbia - Published in 1945, he text features hard to find information on the composition of the gold bearing black sands of the South West Oregon Coast, offering a unique insight to prospectors in search of Oregon's legendary beach gold. 104 ppgs, $9.99

Manganese Mining in Oregon - First released in 1942 and now out of print, this special reprint edition of "Manganese in Oregon" was originally published by the Oregon Department of Geology and Mineral Industries. The text features hard to find information on the mining of Manganese in Oregon, including details and maps of Oregon manganese mines and prospects. 108 ppgs, 9.99

Medford Oregon As A Mining Center - Written in 1912, this hard to find publication includes valuable insights into the mining history of South West Oregon. This small book contains interesting information on the gold, copper and mining industry in Southern Oregon as it existed just prior to World War One, shedding light on some of the important mines in the area. Included are rare photographs and vintage advertising of the day. 80 ppgs, 9.99

Mineral Resources of Curry County Oregon - First released in 1977 and now out of print, this special reprint edition of "Geology, Mineral Resources and Rock Materials of Curry County, Oregon" was originally published in cooperation of Curry County, Oregon and the Oregon Department of Geology and Mineral Industries. The text features hard to find information on not only the mining of gold and other metals in Curry County, but also aggregate mining in the area. 102 ppgs, 11.99

Origin of the Gold Bearing Black Sands of the Coast of South West Oregon - First released in 1943 and now out of print, this special reprint edition of "The Origin of the Black Sands of the South West Oregon Coast" was originally published by the Oregon Department of Geology and Mineral Industries. The text features hard to find information on the origin of the gold bearing black sands of the South West Oregon Coast, offering a unique insight to prospectors in search of Oregon's legendary beach gold. 52 ppgs, 8.99

South West Oregon Mining - Leading mining historian Kerby Jackson introduces us to six classic small mining publications on the Gold Mining Industry in Southern Oregon. This small book consists of a compilation of USGS J.S. Diller's "Mines of the Riddles Quadrangle", "The Rogue River Valley Coal Fields" and "Mineral Resources of the Grants Pass Quadrangle", the Grants Pass Commercial Club's rare publication "Mining in Josephine County, Oregon" and the USGS publication "The Distribution of Placer Gold in the Sixes River, South West Oregon". Also included is F.W. Libbey's legendary article on the Southern Oregon Mining Industry, "Lest We Forget", which appeared in the publication of the Oregon State Department of Geology and Mineral Industries in the early 1960's. This compilation offers a unique perspective on mining in South West Oregon and includes considerable information on mines in Josephine, Jackson and Coos Counties. 142 ppgs, 14.99

Geology and Mineral Resources of the Gasquet Quadrangle of California-Oregon - First published in 1953, it has been unavailable for over a century and sheds important light on the geological features and mineral resources of this portion of Northern California and Southern Oregon. 80 ppgs, 9.99

Idaho Mining Books

Gold in Idaho - Unavailable since the 1940's, this publication was originally compiled by the Idaho Bureau of Mines and includes details on gold mining in Idaho. Included is not only raw data on gold production in Idaho, but also valuable insight into where gold may be found in Idaho, as well as practical information on the gold bearing rocks and other geological features that will assist those looking for placer and lode gold in the State of Idaho. This volume also includes thirteen gold maps that greatly enhance the practical usability of the information contained in this small book detailing where to find gold in Idaho. **8.5" X 11", 72 ppgs. Retail Price: $9.99**

Geology of the Couer D'Alene Mining District of Idaho - Unavailable since 1961, this publication was originally compiled by the Idaho Bureau of Mines and Geology and includes details on the mining of gold, silver and other minerals in the famous Coeur D'Alene Mining District in Northern Idaho. Included are details on the early history of the Coeur D'Alene Mining District, local tectonic settings, ore deposit features, information on the mineral belts of the Osburn Fault, as well as detailed information on the famous Bunker Hill Mine, the Dayrock Mine, Galena Mine, Lucky Friday Mine and the infamous Sunshine Mine. This volume also includes sixteen hard to find maps. **8.5" X 11", 70 ppgs. Retail Price: $9.99**

The Gold Camps and Silver Cities of Idaho - Originally published in 1963, this important publication on Idaho Mining has not been available for nearly fifty years. Included are rare insights into the history of Idaho's Gold Rush, as well as the mad craze for silver in the Idaho Panhandle. Documented in fine detail are the early mining excitements at Boise Basin, at South Boise, in the Owyhees, at Deadwood, Long Valley, Stanley Basin and Robinson Bar, at Atlanta, on the famous Boise River, Volcano, Little Smokey, Banner, Boise Ridge, Hailey, Leesburg, Lemhi, Pearl, at South Mountain, Shoup and Ulysses, Yellow Jacket and Loon Creek. The story follows with the appearance of Chinese miners at the new mining camps on the Snake River, Black Pine, Yankee Fork, Bay Horse, Clayton, Heath, Seven Devils, Gibbonsville, Vienna and Sawtooth City. Also included are special sections on the Idaho Lead and Silver mines of the late 1800's, as well as the mining discoveries of the early 1900's that paved the way for Idaho's modern mining and mineral industry. Lavishly illustrated with rare historic photos, this volume provides a one of a kind documentary into Idaho's mining history that is sure to be enjoyed by not only modern miners and prospectors who still scour the hills in search of nature's treasures, but also those enjoy history and tromping through overgrown ghost towns and long abandoned mining camps. **8.5" X 11", 186 ppgs. Retail Price: $14.99**

Ore Deposits and Mining in North Western Custer County Idaho - Unavailable since 1913, this important publication was originally published by the Us Department of the Interior and has been unavailable for a century. Included are fine details on the geology, geography, gold placers and gold and silver bearing quartz veins of the mining region of North West Custer County, Idaho. Of particular interest is a rare look at the mines and prospects of the region, including those such as the Ramshorn Mine, SkyLark, Riverview, Excelsior, Beardsley, Pacific, Hoosier, Silver Brick, Forest Rose and dozens of others in the Bay Horse Mining District. Also covered are the mines of the Yankee Fork District such as the Lucky Boy, Badger, Black, Enterprise, Charles Dickens, Morrison, Golden Sunbeam, Montana, Golden Gate and others, as well as those in the Loon Mining District. **8.5" X 11", 126 ppgs. Retail Price: $12.99**

Gold Rush To Idaho - Unavailable since 1963, this important publication was originally published by the Idaho Bureau of Mines and has been unavailable for 50 years. "Gold Rush To Idaho" revisits the earliest years of the discovery of gold in Idaho Territory and introduces us to the conditions that the pioneer gold seekers met when they blazed a trail through the wilderness of Idaho's mountains and discovered the precious yellow metal at Oro Fino and Pierce. Subsequent rushes followed at places like Elk City, Newsome, Clearwater Station, Florence, Warrens and elsewhere. Of particular interest is a rare look at the hardships that the first miners in Idaho met with during their day to day existences and their attempts to bring law and order to their mining camps. **8.5" X 11", 88 ppgs. Retail Price: $9.99**

The Geology and Mines of Northern Idaho and North Western Montana - Unavailable since 1909, this important publication was originally published by the Us Department of the Interior and has been unavailable for a century. Included are fine details on the geology and geography of the mining regions of Northern Idaho and North Western Montana. Of particular interest is a rare look at the mines and prospects of the region, including those in the Pine Creek Mining District, Lake Pend Oreille district, Troy Mining District, Sylvanite District, Cabinet Mining District, Prospect Mining District and the Missoula Valley. Some of the mines featured include the Iron Mountain, Silver Butte, Snowshoe, Grouse Mountain Mine and others. **8.5" X 11", 142 ppgs. Retail Price: $12.99**

Mining in the Alturas Quadrangle of Blaine County Idaho - Unavailable since 1922, this important publication was originally published by the Idaho Bureau of Mines and has been unavailable for ninety years. Topics include the geology, rock formations and the formation of ore deposits in this important mining area of Idaho. Of particular focus is information on the local geology, quartz veins and ore deposits of this portion of Idaho. Included are hard to find details, including the descriptions and locations of numerous gold and silver mines in the area including the Silver King, Pilgrim, Columbia, Lone Jack, Sunbeam, Pride of the West, Lucky Boy, Scotia, Atlanta, Beaver-Bidwell and others mines and prospects. **8.5" X 11", 56 ppgs. Retail Price: $8.99**

Mining in Lemhi County Idaho - Originally published in 1913, this important book on Idaho Mining has not been available to miners for over a century. Included are rare insights into hundreds of gold, silver, copper and other mines in this famous Idaho mining area. Details include the locations, geology, history, production and other facts of the mines of this region, not only gold and silver hardrock mines, but also gold placer mines, lead-silver deposits, copper mines, cobalt-nickel deposits, tungsten and tin mines . It is lavishly illustrated with hard to find photos of the period and rare mining maps. Some of the vicinities featured include the Nicholia Mining District, Spring Mountain District, Texas District, Blue Wing District, Junction District, McDevitt District, Pratt Creek, Eldorado District, Kirtley Creek, Carmen Creek, Gibbonsville, Indian Creek, Mineral Hill District, Mackinaw, Eureka District, Blackbird District, YellowJacket District, Gravel Range District, Junction District, Parker Mountain and other mining districts. **8.5" X 11", 226 ppgs. Retail Price: $19.99**

Mining in Shoshone County Idaho - First published in 1923, it has been unavailable for over a century and sheds important light on the mining history of Shoshone County, Idaho. Some of the topics include the history of mining in Shoshone County, a look at the local geology and ore characteristics of lead-silver deposits, zinc deposits, copper, antimony, gold and other minerals. Also included are insights into the history, production, characteristics and locations of numerous mines in the area. 198 ppgs, 15.99

Utah Mining Books

Fluorite in Utah - Unavailable since 1954, this publication was originally compiled by the USGS, State of Utah and U.S. Atomic Energy Commission and details the mining of fluorspar, also known as fluorite in the State of Utah. Included are details on the geology and history of fluorspar (fluorite) mining in Utah, including details on where this unique gem mineral may be found in the State of Utah. 8.5" X 11", 60 ppgs. Retail Price: $8.99

The Gold Hill Mining District of Utah - First published in 1935, it has been unavailable since those days and sheds important light on the mines, history and geology of Utah's Gold Hill Mining District. Included are rare insights into this important mining area, including the locations, histories and details of numerous mines. This volume is well illustrated with geological diagrams, as well as hard to find maps of some of the most important mines in this district. 202 ppgs., 19.99

The Mines, Miners and Minerals of Utah - First published in 1896, it has been unavailable since those days and sheds important light on the early mines and miners of Pioneer Utah, as well as the minerals which they won from the earth by laborious hard physical labor and sheer determination. Included are rare insights into the early mining history of Utah, as well details on hundreds of gold, silver and copper mines. 376 ppgs., 24.99

California Mining Books

The Tertiary Gravels of the Sierra Nevada of California Mining historian Kerby Jackson introduces us to a classic mining work by Waldemar Lindgren in this important re-issue of The Tertiary Gravels of the Sierra Nevada of California. Unavailable since 1911, this publication includes details on the gold bearing ancient river channels of the famous Sierra Nevada region of California. 8.5" X 11", 282 ppgs. Retail Price: $19.99

The Mother Lode Mining Region of California - Unavailable since 1900, this publication includes details on the gold mines of California's famous Mother Lode gold mining area. Included are details on the geology, history and important gold mines of the region, as well as insights into historic mining methods, mine timbering, mining machinery, mining bell signals and other details on how these mines operated. Also included are insights into the gold mines of the California Mother Lode that were in operation during the first sixty years of California's mining history. 8.5" X 11", 176 ppgs. Retail Price: $14.99

Lode Gold of the Klamath Mountains of Northern California and South West Oregon - Unavailable since 1971, this publication was originally compiled by Preston E. Hotz and includes details on the lode mining districts of Oregon and California's Klamath Mountains. Included are details on the geology, history and important lode mines of the French Gulch, Deadwood, Whiskeytown, Shasta, Redding, Muletown, South Fork, Old Diggings, Dog Creek (Delta), Bully Choop (Indian Creek), Harrison Gulch, Hayfork, Minersville, Trinity Center, Canyon Creek, East Fork, New River, Denny, Liberty (Black Bear), Cecilville, Callahan, Yreka, Fort Jones and Happy Camp mining districts in California, as well as the Ashland, Rogue River, Applegate, Illinois River, Takilma, Greenback, Galice, Silver Peak, Myrtle Creek and Mule Creek districts of South Western Oregon. Also included are insights into the mineralization and other characteristics of this important mining region. 8.5" X 11", 100 ppgs. Retail Price: $10.99

Mines and Mineral Resources of Shasta County, Siskiyou County, Trinity County: California - Unavailable since 1915, this publication was originally compiled by the California State Mining Bureau and includes details on the gold mines of this area of Northern California. Also included are insights into the mineralization and other characteristics of this important mining region, as well as the location of historic gold mines. 8.5" X 11", 204 ppgs. Retail Price: $19.99

Geology of the Yreka Quadrangle, Siskiyou County, California - Unavailable since 1977, this publication was originally compiled by Preston E. Hotz and includes details on the geology of the Yreka Quadrangle of Siskiyou County, California. Also included are insights into the mineralization and other characteristics of this important mining region. **8.5" X 11", 78 ppgs. Retail Price: $7.99**

Mines of San Diego and Imperial Counties, California - Originally published in 1914, this important publication on California Mining has not been available for a century. This publication includes important information on the early gold mines of San Diego and Imperial County, which were some of the first gold fields mined in California by early Spanish and Mexican miners before the 49ers came on the scene. Included are not only details on early mining methods in the area, production statistics and geological information, but also the location of the early gold mines that helped make California "The Golden State". Also included are details on the mining of other minerals such as silver, lead, zinc, manganese, tungsten, vanadium, asbestos, barite, borax, cement, clay, dolomite, fluospar, gem stones, graphite, marble, salines, petroleum, stronium, talc and others. **8.5" X 11", 116 ppgs. Retail Price: $12.99**

Mines of Sierra County, California - Unavailable since 1920, this publication was originally compiled by the California State Mining Bureau and includes details on the gold mines of Sierra County, California. Also included are insights into the mineralization and other characteristics of this important mining region, as well as the location of historic gold mines. **8.5" X 11", 156 ppgs. Retail Price: $19.99**

Mines of Plumas County, California - Unavailable since 1918, this publication was originally compiled by the California State Mining Bureau and includes details on the gold mines of Plumas County, California. Also included are insights into the mineralization and other characteristics of this important mining region, as well as the location of historic gold mines. **8.5" X 11", 200 ppgs. Retail Price: $19.99**

Mines of El Dorado, Placer, Sacramento and Yuba Counties, California - Originally published in 1917, this important publication on California Mining has not been available for nearly a century. This publication includes important information on the early gold mines of El Dorado County, Placer County, Sacramento County and Yuba County, which were some of the first gold fields mined by the Forty-Niners during the California Gold Rush. Included are not only details on early mining methods in the area, production statistics and geological information, but also the location of the early gold mines that helped make California "The Golden State". Also included are insights into the early mining of chrome, copper and other minerals in this important mining area. **8.5" X 11", 204 ppgs. Retail Price: $19.99**

Mines of Los Angeles, Orange and Riverside Counties, California - Originally published in 1917, this important publication on California Mining has not been available for nearly a century. This publication includes important information on the early gold mines of Los Angeles County, Orange County and Riverside County, which were some of the first gold fields mined in California by early Spanish and Mexican miners before the 49ers came on the scene. Included are not only details on early mining methods in the area, production statistics and geological information, but also the location of the early gold mines that helped make California "The Golden State". **8.5" X 11", 146 ppgs. Retail Price: $12.99**

Mines of San Bernadino and Tulare Counties, California - Originally published in 1917, this important publication on California Mining has not been available for nearly a century. This publication includes important information on the early gold mines of San Bernadino and Tulare County, which were some of the first gold fields mined in California by early Spanish and Mexican miners before the 49ers came on the scene. Included are not only details on early mining methods in the area, production statistics and geological information, but also the location of the early gold mines that helped make California "The Golden State". Also included are details on the mining of other minerals such as copper, iron, lead, zinc, manganese, tungsten, vanadium, asbestos, barite, borax, cement, clay, dolomite, fluospar, gem stones, graphite, marble, salines, petroleum, stronium, talc and others. **8.5" X 11", 200 ppgs. Retail Price: $19.99**

Chromite Mining in The Klamath Mountains of California and Oregon - Unavailable since 1919, this publication was originally compiled by J.S. Diller of the United States Department of Geological Survey and includes details on the chromite mines of this area of Northern California and Southern Oregon. Also included are insights into the mineralization and other characteristics of this important mining region, as well as the location of historic mines. Also included are insights into chromite mining in Eastern Oregon and Montana. **8.5" X 11", 98 ppgs. Retail Price: $9.99**

Mines and Mining in Amador, Calaveras and Tuolumne Counties, California - Unavailable since 1915, this publication was originally compiled by William Tucker and includes details on the mines and mineral resources of this important California mining area. Included are details on the geology, history and important gold mines of the region, as well as insights into other local mineral resources such as asbestos, clay, copper, talc, limestone and others. Also included are insights into the mineralization and other characteristics of this important portion of California's Mother Lode mining region. **8.5" X 11", 198 ppgs. Retail Price: $14.99**

The Cerro Gordo Mining District of Inyo County California - Unavailable since 1963, this publication was originally compiled by the United States Department of Interior. Included are insights into the mineralization and other characteristics of this important mining region of Southern California. Topics include the mining of gold and silver in this important mining district in Inyo County, California, including details on the history, production and locations of the Cerro Gordo Mine, the Morning Star Mine, Estelle Tunnel, Charles Lease Tunnel, Ignacio, Hart, Crosscut Tunnel, Sunset, Upper Newtown, Newtown, Ella, Perseverance, Newsboy, Belmont and other silver and gold mines in the Cerro Gordo Mining District. This volume also includes important insights into the fossil record, geologic formations, faults and other aspects of economic geology in this California mining district. 8.5" X 11", 104 ppgs. Retail Price: $10.99

Mining in Butte, Lassen, Modoc, Sutter and Tehama Counties of California - Unavailable since 1917, this publication was originally compiled by the United States Department of Interior. Included are insights into the mineralization and other characteristics of this important mining region of California. Topics include the mining of asbestos, chromite, gold, diamonds and manganese in Butte County, the mining of gold and copper in the Hayden Hill and Diamond Mountain mining districts of Lassen County, the mining of coal, salt, copper and gold in the High Grade and Winters mining districts of Modoc County, gold mining in Sutter County and the mining of gold, chromite, manganese and copper in Tehama County. This volume also includes the production records and locations of numerous mines in this important mining region. 8.5" X 11", 114 ppgs. Retail Price: $11.99

Mines of Trinity County California - Originally published in 1965, this important publication on California Mining has not been available for nearly fifty years. This publication includes important information on mines and mining in Trinity County, California, as well insights into the mineralization and geology of this important mining area in Northern California. Included are extensive details on hardrock and placer gold mines and prospects, including charts showing the locations of these historic mines.. 8.5" X 11", 144 ppgs. Retail Price: $12.99

Mines of Kern County California - Originally published in 1962, this important publication on California Mining has not been available for nearly fifty years. This publication includes important information on mines and mining in Kern County, California, as well insights into the mineralization and geology of this important mining area in California. Included are extensive details on hardrock and placer gold mines and prospects, including charts showing the locations of these historic mines. 8.5" X 11", 398 ppgs. Retail Price: $24.99

Mines of Calaveras County California - Originally published in 1962, this important publication on California Mining has not been available for nearly fifty years. This publication includes important information on mines and mining in Calaveras County, California, as well insights into the mineralization and geology of this important mining area in Northern California. Included are extensive details on hardrock and placer gold mines and prospects, including charts showing the locations of these historic mines. 8.5" X 11", 236 ppgs. Retail Price: $19.99

Lode Gold Mining in Grass Valley California - Unavailable since 1940, this publication was originally compiled by the United States Department of Interior. Included are insights into the gold mineralization and other characteristics of this important mining region of Nevada County, California. This volume also includes important insights into the geologic formations, faults and other aspects of economic geology in this California mining district. Of particular interest are the fine details on many hardrock gold mines in the area, including their locations, histories, development and mineralization. Some of the mines featured include the Gold Hill Mine, Massachusetts Hill, Boundary, Peabody, Golden Center, North Star, Omaha, Lone Jack, Homeward Bound, Hartery, Wisconsin, Allison Ranch, Phoenix, Kate Hayes, W.Y.O.D., Empire, Rich Hill, Daisy Hill, Orleans, Sultana, Centennial, Conlin, Ben Franklin, Crown Point and many others. 8.5" X 11", 148 ppgs. Retail Price: $12.99

Lode Mining in the Alleghany District of Sierra County California - Unavailable since 1913, this publication was originally compiled by the United States Department of Interior. Included are insights into the mineralization and other characteristics of this important mining region of Sierra County. Included are details on the history, production and locations of numerous hardrock gold mines in this famous California area, including the Tightner Mine, Minnie D., Osceola, Eldorado, Twenty One, Sherman, Kenton, Oriental, Rainbow, Plumbago, Irelan, Gold Canyon, North Fork, Federal, Kate Hardy and others. This volume also includes important insights into the fossil record, geologic formations, faults and other aspects of economic geology in this California mining district. 8.5" X 11", 48 ppgs. Retail Price: $7.99

Six Months In The Gold Mines During The California Gold Rush - Unavailable since 1850, this important work is a first hand account of one "49'ers" personal experience during the great California Gold Rush, shedding important light on one of the most exciting periods in the history of not only California, but also the world. Compiled from journals written between 1847 and 1849 by E. Gould Buffum, a native of New York, "Six Months In The Gold Mines During The California Gold Rush" offers a rare look into the day to day lives of the people who came to California to work in her gold mines when the state was still a great frontier. 8.5" X 11", 290 ppgs. Retail Price: $19.99

Quartz Mines of the Grass Valley Mining District of California - Unavailable since 1867, this important publication has not been available since those days. This rare publication offers a short dissertation on the early hardrock mines in this important mining district in the California Mother Lode region between the 1850's and 1860's. Also included are hard to find details on the mineralization and locations of these mines, as well as how they were operated in those day. **8.5" X 11", 44 ppgs. Retail Price: $8.99**

Gold Rush on the Feather River - First published in 1924, this short publication by G.C. Mansfield sheds important light on the early history of gold mining on the Feather River. Included are rare insights into the first decade of gold mining and the early mining camps of the Feather River during the 1850's. 64 ppgs., 9.99

The Bodie Mining District of California - First published in 1986, it has been unavailable since those days and sheds important light on this famous mining area. Included are the history, characteristics and locations of numerous old mines around the ghost town of Bodie. 64 ppgs, 8.99

Geology and Mineral Resources of the Gasquet Quadrangle of California-Oregon - First published in 1953, it has been unavailable for over a century and sheds important light on the geological features and mineral resources of this portion of Northern California and Southern Oregon. 80 ppgs, 9.99

Alaska Mining Books

Ore Deposits of the Willow Creek Mining District, Alaska - Unavailable since 1954, this hard to find publication includes valuable insights into the Willow Creek Mining District near Hatcher Pass in Alaska. The publication includes insights into the history, geology and locations of the well known mines in the area, including the Gold Cord, Independence, Fern, Mabel, Lonesome, Snowbird, Schroff-O'Neil, High Grade, Marion Twin, Thorpe, Webfoot, Kelly-Willow, Lane, Holland and others. **8.5" X 11", 96 ppgs. Retail Price: $9.99**

The Juneau Gold Belt of Alaska - Unavailable since 1906, this hard to find publication includes valuable insights into the gold mines around Juneau, Alaska. The publication includes important details into the history, geology and locations of the well known gold mines and prospects in the area, including those around Windham Bay, Holkham Bay, Port Snettisham, on Grindstone and Rhine Creeks, Gold Creek, Douglas Island, Salmon Creek, Lemon Creek, Nugget Creek, from the Mendenhall River to Berners Bay, McGinnis Creek, Montana Creek, Peterson Creek, Windfall Creek, the Eagle River, Yankee Basin, Yankee Curve, Kowee Creek and elsewhere. Not only are gold placer mines included, but also hardrock gold mines. **8.5" X 11", 224 ppgs. Retail Price: $19.99**

Mining in the Jumbo Basin of Alaska - Unavailable since 1953, this hard to find publication includes valuable insights into the mines and geology of the Jumbo Basin. The publication includes important details into the history, geology and locations of the well known gold mines and prospects in the famous Jumbo Basin Mining Region of Alaska. 72 ppgs, 9.99

The Rampart Placer Gold Region of Alaska - Unavailable since 1906, this hard to find publication includes valuable insights into the placer gold mines of the Rampart Mining Region. The publication includes important details into the history, geology and locations of the well known gold mines and prospects in the famous Rampart Mining Region of Alaska. 78 ppgs, 10.99

Arizona Mining Books

Mines and Mining in Northern Yuma County Arizona - Originally published in 1911, this important publication on Arizona Mining has not been available for over a hundred years. Included are rare insights into the gold, silver, copper and quicksilver mines of Yuma County, Arizona together with hard to find maps and photographs. Some of the mines and mining districts featured include the Planet Copper Mine, Mineral Hill, the Clara Consolidated Mine, Viati Mine, Copper Basin prospect, Bowman Mine, Quartz King, Billy Mack, Carnation, the Wardwell and Osbourne, Valensuella Copper, the Mariquita, Colonial Mine, the French American, the New York-Plomosa, Guadalupe, Lead Camp, Mudersbach Copper Camp, Yellow Bird, the Arizona Northern (Salome Strike), Bonanza (Harqua Hala), Golden Eagle, Hercules, Socorro and others. **8.5" X 11", 144 ppgs. Retail Price: $11.99**

The Aravaipa and Stanley Mining Districts of Graham County Arizona - Originally published in 1925, this important publication on Arizona Mining has not been available for nearly ninety years. Included are rare insights into the gold and silver mines of these two important mining districts, together with hard to find maps. **8.5" X 11", 140 ppgs. Retail Price: $11.99**

Gold in the Gold Basin and Lost Basin Mining Districts of Mohave County, Arizona - This volume contains rare insights into the geology and gold mineralization of the Gold Basin and Lost Basin Mining Districts of Mohave County, Arizona that will be of benefit to miners and prospectors. Also included is a significant body of information on the gold mines and prospects of this portion of Arizona. This volume is lavishly illustrated with rare photos and mining maps. 8.5" X 11", 188 ppgs. Retail Price: $19.99

Mines of the Jerome and Bradshaw Mountains of Arizona - This important publication on Arizona Mining has not been available for ninety years. This volume contains rare insights into the geology and ore deposits of the Jerome and Bradshaw Mountains of Arizona that will be of benefit to miners and prospectors who work those areas. Included is a significant body of information on the mines and prospects of the Verde, Black Hills, Cherry Creek, Prescott, Walker, Groom Creek, Hassayampa, Bigbug, Turkey Creek, Agua Fria, Black Canyon, Peck, Tiger, Pine Grove, Bradshaw, Tintop, Humbug and Castle Creek Mining Districts. This volume is lavishly illustrated with rare photos and mining maps. 8.5" X 11", 218 ppgs. Retail Price: $19.99

The Ajo Mining District of Pima County Arizona - This important publication on Arizona Mining has not been available for nearly seventy years. This volume contains rare insights into the geology and mineralization of the Ajo Mining District in Pima County, Arizona and in particular the famous New Cornelia Mine. 8.5" X 11", 126 ppgs. Retail Price: $11.99

Mining in the Santa Rita and Patagonia Mountains of Arizona - Originally published in 1915, this important publication on Arizona Mining has not been available for nearly a century. Included are rare insights into hundreds of gold, silver, copper and other mines in this famous Arizona mining area. Details include the locations, geology, history, production and other facts of the mines of this region. 8.5" X 11", 394 ppgs. Retail Price: $24.99

Mining in the Bisbee Quadrangle of Arizona - Originally published in 1906, this important publication on Arizona Mining has not been available for nearly a century. Included are rare insights into hundreds of gold, silver, copper and other mines in this famous Arizona mining area. Details include the locations, geology, history, production and other facts of the mines of this important mining region. 8.5" X 11", 188 ppgs. Retail Price: $14.99

Placer Gold Mining in Arizona - Unavailable since 1922, this hard to find publication includes valuable insights into the placer gold mines of the Arizona. Originally released as "Placer Gold of Arizona", despite its small size, this publication includes important details into the history, geology and locations of the well known placer gold mines and prospects in the State of Arizona. 48 ppgs, 8.99

Gold and Copper Mining near Payson, Arizona - Written in 1915, this hard to find publication includes valuable insights into the gold and copper mining industry of Arizona. Highlighted here are the gold and copper mines near Payson, Arizona. 68 ppgs, 8.99

Lode Gold Mining in Arizona - Unavailable since 1934, this hard to find publication, originally released as "Arizona Lode Gold Mines and Gold Mining" includes valuable insights into the gold mining industry of Arizona. Included are valuable insights into over 150 hardrock gold mines in over 30 different mining districts in Arizona. 278 ppgs, 21.99

Mining in the Dragoon Quadrangle of Cochise County, Arizona - Unavailable since 1964, this hard to find publication includes valuable insights into the mines of the Dragoon Quadrangle Mining Region. The publication includes important details into the history, geology and locations of the well known mines and prospects in this famous mining region of Arizona. 224 ppgs., 19.99

Directory of Operating Mines in Arizona in 1915 - Unavailable since 1916, this hard to find publication includes valuable insights into the mines of Arizona. This small publication includes a complete list of the mines that were operating in the State of Arizona during 1915 and includes details such as general location, owners and some basic facts about each mining operation.52 ppgs. 8.99

Arizona Ore Deposits - Unavailable since 1938, this hard to find publication includes valuable insights into some ore deposits of Arizona. Included are valuable insights into the formation and characteristics of valuable ore deposits in the Jerome, Miami, Inspiration, Clifton, Morenci, Ray, Ajo, Eureka, Tombstone and Magma mining districts. Included are details into some of the major gold, silver and copper mines of these important Arizona mining areas. 160 ppgs, 14.99

Montana Mining Books

A History of Butte Montana: The World's Greatest Mining Camp - First published in 1900 by H.C. Freeman, this important publication sheds a bright light on one of the most important mining areas in the history of The West. Together with his insights, as well as rare photographs of the periods, Harry Freeman describes Butte and its vicinity from its early beginnings, right up to its flush years when copper flowed from its mines like a river. At the time of publication, Butte, Montana was known worldwide as "The Richest Mining Spot On Earth" and produced not only vast amounts of copper, but also silver, gold and other metals from its mines. Freeman illustrates, with great detail, the most important mines in the vicinity of Butte, providing rare details on their owners, their history and most importantly, how the mines operated and how their treasures were extracted. Of particular interest are the dozens of rare photographs that depict mines such as the famous Anaconda, the Silver Bow, the Smoke House, Moose, Paulin, Buffalo, Little Minah, the Mountain Consolidated, West Greyrock, Cora, the Green Mountain, Diamond, Bell, Parnell, the Neversweat, Nipper, Original and many others. **8.5" X 11", 142 ppgs. Retail Price: $12.99**

The Butte Mining District of Montana - This important publication on Montana Mining has not been available for over a century. Included are rare insights into the gold, copper and silver mines of Butte, Montana together with hard to find maps and photographs. Some of the topics include the early history of gold, silver and copper mining in the Butte area, insight into the geology of its mining areas, the local distribution of gold, silver and copper ores, as well their composition and how to identify them. Also included are detailed facts about the mines in the Butte Mining District, including the famous Anaconda Mine, Gagnon, Parrot, Blue Vein, Moscow, Poulin, Stella, Buffalo, Green Mountain, Wake Up Jim, the Diamond-Bell Group, Mountain Consolidated, East Greyrock, West Greyrock, Snowball, Corra, Speculator, Adirondack, Miners Union, the Jessie-Edith May Group, Otisco, Iduna, Colorado, Lizzie, Cambers, Anderson, Hesperus, Preferencia and dozens of others. **8.5" X 11", 298 ppgs. Retail Price: $24.99**

Mines of the Helena Mining Region of Montana - This important publication on Montana Mining has not been available for over a century. Included are rare insights into the gold, copper and silver mines of the vicinity of Helena, Montana, including the Marysville Mining District, Elliston Mining District, Rimini Mining District, Helena Mining District, Clancy Mining District, Wickes Mining District, Boulder and Basin Mining Districts and the Elkhorn Mining District. Some of the topics include the early history of gold, silver and copper mining in the Helena area, insight into the geology of its mining areas, the local distribution of gold, silver and copper ores, as well their composition and how to identify them. Also included are detailed facts, history, geology and locations of over one hundred gold, silver and copper mines in the area . **8.5" X 11", 162 ppgs, Retail Price: $14.99**

Mines and Geology of the Garnet Range of Montana - This important publication on Montana Mining has not been available for over a century. Included are rare insights into the gold, copper and silver mines of the vicinity of this important mining area of Montana. Some of the topics include the early history of gold, silver and copper mining in the Garnet Mountains, insight into the geology of its mining areas, the local distribution of gold, silver and copper ores, as well their composition and how to identify them. Also included are detailed facts, history, geology and locations of numerous gold, silver and copper mines in the area . **8.5" X 11", 100 ppgs, Retail Price: $11.99**

Mines and Geology of the Philipsburg Quadrangle of Montana - This important publication on Montana Mining has not been available for over a century. Included are rare insights into the gold, copper and silver mines of the vicinity of this important mining area of Montana. Some of the topics include the early history of gold, silver and copper mining in the Philipsburg Quadrangle, insight into the geology of its mining areas, the local distribution of gold, silver and copper ores, as well their composition and how to identify them. Also included are detailed facts, history, geology and locations of over one hundred gold, silver and copper mines in the area **8.5" X 11", 290 ppgs, Retail Price: $24.99**

Geology of the Marysville Mining District of Montana - Included are rare insights into the mining geology of the Marysville Mining District. Some of the topics include the early history of gold, silver and copper mining in the area, insight into the geology of its mining areas, the local distribution of gold, silver and copper ores, as well their composition and how to identify them. Also included are detailed facts, history, geology and locations of gold, silver and copper mines in the area **8.5" X 11", 198 ppgs, Retail Price: $19.99**

The Geology and Mines of Northern Idaho and North Western Montana - See listing under Idaho.

The History of Gold Dredging in Montana - Unavailable since 1916, this important publication was originally published by the Us Bureau of Mines and has been unavailable for a century. A century and more ago, giant dredging machines dug in Montana's rivers and creeks in search of illusive golden riches. First appearing in California in the 1850's, gold dredges finally reached their peak of development in Siberia and New Zealand before becoming popular again in the United States. This book offers a unique historical perspective on the gold dredges that once operated in Montana. This book on Montana mining history is lavishly illustrated with dozens of rare historic photos gold dredges that once operated in Montana, as well as hard to locate plans on how these dredges were designed. 120 ppgs., 11.99

Nevada Mining Books

The Bull Frog Mining District of Nevada - Unavailable since 1910, this publication was originally compiled by the United States Department of Interior. This volume also includes important insights into the geologic formations, faults and other aspects of economic geology in this Nevada mining district. Of particular interest are the fine details on many mines in the area, including their locations, histories, development and mineralization. Some of the mines featured include the National Bank Mine, Providence, Gibraltor, Tramps, Denver, Original Bullfrog, Gold Bar, Mayflower, Homestake-King and other mines and prospects. **8.5" X 11", 152 ppgs, Retail Price: $14.99**

History of the Comstock Lode - Unavailable since 1876, this publication was originally released by John Wiley & Sons. This volume also includes important insights into the famous Comstock Lode of Nevada that represented the first major silver discovery in the United States. During its spectacular run, the Comstock produced over 192 million ounces of silver and 8.2 million ounces of gold. Not only did the Comstock result in one of the largest mining rushes in history and yield immense fortunes for its owners, but it made important contributions to the development of the State of Nevada, as well as neighboring California. Included here are important details on not only the early development and history of the Comstock, but also rare early insight into its mines, ore and its geology.**8.5" X 11", 244 ppgs, Retail Price: $19.99**

The Pioche Mining District of Nevada - First published in 1932, it has been unavailable for over a century and sheds important light on the mining history of Nevada. Some of the topics include the history of mining in this district, as well as the characteristics of its mineral and ore deposits. Also included are insights into the history, production, characteristics and locations of numerous mines in the area. Some of the mines include the Combined Metals, Pioche, Ely Valley, No. 10, Poorman, Wide Awake, Alps, Prince, Virginia Louise, Half Moon, Abe Lincoln, Fairview, Bristol Silver, National, Vesuvius, Inman, Tempest, Hillside, Jackrabbit, Lucky Star, Fortuna, Mendha, Manhattan, Hamburg, Comet, Lyndon and others. 108 ppgs 10.99

The Yerington Mining District of Nevada - First published in 1932, it has been unavailable for over a century and sheds important light on the mining history of Nevada. Some of the topics include the history of mining in this district, as well as the characteristics of its mineral and ore deposits. Also included are insights into the history, production, characteristics and locations of numerous mines in the area. Some of the mines include the Bluestone, Mason Valley, Malachite, McConnell, Greenwood, Western Nevada, Ludwig, Douglas Hill, Casting Copper, Montana-Yerington, Empire, Jim Beatty, Terry and McFarland, Blue Jay and others. 92 ppgs, 10.99

The Genesis of the Ores of Tonopah Nevada - Unavailable since 1918, this hard to find publication includes valuable insights into the gold mines around Tonopah, Nevada. The publication includes important details into the geology of mines in the Tonopah Mining District of Nevada. 90 ppgs, 10.99

Mining Camps of Elko, Lander and Eureka Counties Nevada - Unavailable since 1910, this hard to find publication includes valuable insights into the mining camps of Elko, Lander and Eureka Counties, Nevada. The publication includes important details into the history of mines and mining in these three Nevada counties. 154 ppgs, 12.99

Ore Deposits of the Bullfrog Quadrangle - Unavailable since 1964 and released as "Geology of Bullfrog Quadrangle and Ore Deposits Related to Bullfrog Hills Caldera, Nye County, Nevada and Inyo County, California". The publication includes important details into the geology of mines in the Bullfrog Quadrangle of Nye County, Nevada and Inyo County, California. 52 ppgs, 9.99

Mining in Eureka County Nevada - Unavailable since 1879, this hard to find publication includes valuable insights into the early mining history off Eureka County, Nevada. The publication includes important details into the early history of the mines of Eureka County, as well as their development, production and how their ores were treated. Also included are details on the 1872 Mining Act, as well as the local rules, regulations and customs of the miners in Eureka County.134 ppgs, 12.99

Colorado Mining Books

Ores of The Leadville Mining District - Unavailable since 1926, this publication was originally compiled by the United States Department of Interior. This volume also includes important insights into the ores and mineralization of the Leadville Mining District in Colorado. Topics include historic ore prospecting methods, local geology, insights into ore veins and stockworks, the local trend and distribution of ore channels, reverse faults, shattered rock above replacement ore bodies, mineral enrichment in oxidized and sulphide zones and more. **8.5" X 11", 66 ppgs, Retail Price: $8.99**

Mining in Colorado - Unavailable since 1926, this publication was originally compiled by the United States Department of Interior. This volume also includes important insights into the mining history of Colorado from its early beginnings in the 1850's right up to the mid 1920's. Not only is Colorado's gold mining heritage included, but also its silver, copper, lead and zinc mining industry. Each mining area is treated separately, detailing the development of Colorado's mines on a county by county basis. **8.5" X 11", 284 ppgs, Retail Price: $19.99**

Gold Mining in Gilpin County Colorado - Unavailable since 1876, this publication was originally compiled by the Register Steam Printing House of Central City, Colorado. A rare glimpse at the gold mining history and early mines of Gilpin County, Colorado from their first discovery in the 1850's up to the "flush years" of the mid 1870's. Of particular interest is the history of the discovery of gold in Gilpin County and details about the men who made those first strikes. Special focus is given to the early gold mines and first mining districts of the area, many of which are not detailed in other books on Colorado's gold mining history. **8.5" X 11", 156 ppgs, Retail Price: $12.99**

Mining in the Gold Brick Mining District of Colorado - Important insights into the history of the Gold Brick Mining District, as well as its local geography and economic geology. Also included are the histories and locations of historic mines in this important Colorado Mining District, including the Cortland, Carter, Raymond, Gold Links, Sacramento, Bassick, Sandy Hook, Chronicle, Grand Prize, Chloride, Granite Mountain, Lucille, Gray Mountain, Hilltop, Maggie Mitchell, Silver Islet, Revenue, Roosevelt, Carbonate King and others. In addition to hardrock mining, are also included are details on gold placer mining in this portion of Colorado. **8.5" X 11", 140 ppgs, Retail Price: $12.99**

Ore Deposits of the London Fault of Colorado - First published in 1941, it has been unavailable since those days and sheds important light on the mines and mineral deposits of the London Fault in Central Colorado's Alma Mining District. This publication sheds important light on the gold veins and lead-silver deposits of the Alma Mining District. Included are geologic details on the London Mine, American Mine, Havigorst Tunnel, Ophir Mine, Mosher Tunnel, London-Butte Mine, Venture Shaft, Hard-To-Beat Mine, Oliver Twist Tunnel, Sacramento Mine, Mudsill Mine, Sherwood Mine, Wagner, Barcoe Tunnel and other mines in this important mining region. 110 ppgs., 10.99

The Mines of Colorado - First published in 1867, it has been unavailable since those days and sheds important light on Colorado's early mining history. Written shortly after the events took place, this publication sheds important light on the Pike's Peak Gold Rush, the discovery of gold on Ralston Creek and Dry Creek in the 1850's, as well as details on the first wave of miners into Colorado and their trials and tribulations as they crossed the Great Plains. Also included are details on early discoveries of lode gold in the mountainous regions of Colorado, details on the early mines hardrock and placer mines, and much more. It is a veritable treasure trove on Colorado's early mining history and will be of great importance to anyone who is interested in the mining of gold or other minerals in Colorado, as well as those interested in the history of the state. 478 ppgs., 29.99

The La Plata Mining District of Colorado - Originally titled "Geology and Ore Deposits in the Vicinity of the La Plata District of Colorado" and first published in 1949, it has been unavailable since those days and sheds important light on the mines and mineral deposits of the La Plata Mining District of Colorado. 214 ppgs., 19.99

Washington Mining Books

The Republic Mining District of Washington - Unavailable since 1910, this important publication was originally published by the Washington Geologic Survey and has been unavailable for a century. Topics include the geology, rock formations and the formation of ore deposits in this important mining area of Washington State. Also included are hard to find details on the geology, history and locations of dozens of mines in the area. Some of the mines featured include the New Republic Mine, Ben Hur, Morning Glory, the South Republic Mine, Quilp, Surprise, Black Tail, Lone Pine, San Poil, Mountain Lion, Tom Thumb, Elcaliph and many others. **8.5" X 11", 94 ppgs, Retail Price: $10.99**

The Myers Creek and Nighthawk Mining Districts of Washington - Unavailable since 1911, this important publication was originally published by the Washington Geologic Survey and has been unavailable for a century. Topics include the geology, rock formations and the formation of ore deposits in these important mining areas of Washington State. Also included are hard to find details on the geology, history and locations of dozens of mines in the area. Some of the mines featured include the Grant Mine, Monterey, Nip and Tuck, Myers Creek, Number Nine, Neutral, Rainbow, Aztec, Crystal Butte, Apex, Butcher Boy, Molson, Mad River, Olentangy, Delate, Kelsey, Golden Chariot, Okanogan, Ohio, Forty-Ninth Parallel, Nighthawk, Favorite, Little Chopaka, Summit, Number One, California, Peerless, Caaba, Prize Group, Ruby, Mountain Sheep, Golden Zone, Rich Bar, Similkameen, Kimberly, Triune, Hiawatha, Trinity, Hornsilver, Maquae, Bellevue, Bullfrog, Palmer Lake, Ivanhoe, Copper World and many others. **8.5" X 11", 136 ppgs, Retail Price: $12.99**

The Blewett Mining District of Washington - Unavailable since 1911, this important publication was originally published by the Washington Geologic Survey and has been unavailable for a century. Topics include the geology, rock formations and the formation of ore deposits in this important mining area of Washington State. Also included are hard to find details on the geology, history and locations of dozens of mines in the area. Some of the mines featured include the Washington Meteor, Alta Vista, Pole Pick, Blinn, North Star, Golden Eagle, Tip Top, Wilder, Golden Guinea, Lucky Queen, Blue Bell, Prospect, Homestake, Lone Rock, Johnson, and others. **8.5" X 11", 134 ppgs, Retail Price: $12.99**

Silver Mining In Washington - Unavailable since 1955, this important publication was originally published by the Washington Geologic Survey. Featured are the hard to find locations and details pertaining to Washington's silver mines. **8.5" X 11", 180 ppgs, Retail Price: $15.99**

The Mines of Snohomish County Washington - Unavailable since 1942, this important publication was originally published by the Washington Geologic Survey and has been unavailable for seventy years. Featured are details on a large number of gold, silver, copper, lead and other metallic mineral mines. Included are the locations of each historic mine, along with information on the commodity produced. **8.5" X 11", 98 ppgs, Retail Price: $10.99**

The Mines of Chelan County Washington - Unavailable since 1943, this important publication was originally published by the Washington Geologic Survey and has been unavailable for seventy years. Featured are details on a large number of gold, silver, copper, lead and other metallic mineral mines. Included are the locations of each historic mine, along with information on the commodity. **8.5" X 11", 88 ppgs, Retail Price: $9.99**

Metal Mines of Washington - Unavailable since 1921, this important publication was originally published by the Washington Geologic Survey and has been unavailable for nearly ninety years. Widely considered a masterpiece on the Washington Mining Industry, "Metal Mines of Washington" sheds light on the important details of Washington's early mining years. Featured are details on hundreds of gold, silver, copper, lead and other metallic mineral mines. Included are hard to find details on the mineral resources of this state, as well as the locations of historic mines. Lavishly illustrated with maps and historic photos and complete with a glossary to explain any technical terms found in the text, this is one of the most important works on mining in the State of Washington. No prospector or miner should be without it if they are interested in mining in Washington. **8.5" X 11", 396 ppgs, Retail Price: $24.99**

Gem Stones In Washington - Unavailable since 1949, this important publication was originally published by the Washington Geologic Survey and has been unavailable since first published. Included are details on where to find naturally occurring gem stones in the State of Washington, including quartz crystal, amethyst, smoky quartz, milky quartz, agates, bloodstone, carnelian, chert, flint, jasper, onyx, petrified wood, opal, fire opal, hyalite and others. **8.5" X 11", 54 ppgs, Retail Price: $8.99**

The Covada Mining District of Washington - Unavailable since 1913, this important publication was originally published by the Washington Geologic Survey and has been unavailable for a century. Topics include the geology, rock formations and the formation of ore deposits in this important mining area of Washington State. Also included are hard to find details on the geology, history and locations of dozens of mines in the area. Some of the mines featured include the Admiral, Advance, Algonkian, Big Bug, Big Chief, Big Joker, Black Hawk, Black Tail, Black Thorn, Captain, Cherokee Strip, Colorado, Dan Patch, Dead Shot, Etta, Good Ore, Greasy Run, Great Scott, Idora, IXL, Jay Bird, Kentucky Bell, King Solomon, Laurel, Laura S, Little Jay, Meteor, Neglected, Northern Light, Old Nell, Plymouth Rock, Polaris, Quandary, Reserve, Shoo Fly, Silver Plume, Three Pines, Vernie, White Rose and dozens of others. **8.5" X 11", 114 ppgs, Retail Price: $10.99**

The Index Mining District of Washington - Unavailable since 1912, this important publication was originally published by the Washington Geologic Survey and has been unavailable for a century. Topics include the geology, rock formations and the formation of ore deposits in this important mining area of Washington State. Also included are hard to find details on the geology, history and locations of dozens of mines in the area. Some of the mines featured include the Sunset, Non-Pareil, Ethel Consolidated, Kittaning, Merchant, Homestead, Co-operative, Lost Creek, Uncle Sam, Calumet, Florence-Rae, Bitter Creek, Index Peacock, Gunn Peak, Helena, North Star, Buckeye. Copper Bell, Red Cross and others. **8.5" X 11", 114 ppgs, Retail Price: $11.99**

Mining & Mineral Resources of Stevens County Washington - Unavailable since 1920, this important publication was originally published by the Washington Geologic Survey and has been unavailable for a century. Topics include the geology, rock formations and the formation of ore deposits in these important mining areas of Washington State. Also included are hard to find details on the geology, history and locations of hundreds of mines in the area. **8.5" X 11", 372 ppgs, Retail Price: $24.99**

The Mines and Geology of the Loomis Quadrangle Okanogan County, Washington - Unavailable since 1972, this important publication was originally published by the Washington Geologic Survey and has been unavailable for a century. Topics include the geology, rock formations and the formation of ore deposits in this important mining area of Washington State. Also included are hard to find details on the geology, history and locations of dozens of gold, copper, silver and other mines in the area. **8.5" X 11", 150 ppgs, Retail Price: $12.99**

The Conconully Mining District of Okanogan County Washington - Unavailable since 1973, this important publication was originally published by the Washington Geologic Survey and has been unavailable for a century. Topics include the geology, rock formations and the formation of ore deposits in this important mining area of Washington State, which also includes Salmon Creek, Blue Lake and Galena. Also included are hard to find details on the geology, mining history and locations of dozens of mines in the area. Some of the mines include Arlington, Fourth of July, Sonny Boy, First Thought, Last Chance, War Eagle-Peacock, Wheeler, Mohawk, Lone Star, Woo Loo Moo Loo, Keystone, Hughes, Plant-Callahan, Johnny Boy, Leuena, Gubser, John Arthur, Tough Nut, Homestake, Key and many others **8.5" X 11", 68 ppgs, Retail Price: $8.99**

Wyoming Mining Books

Mining in the Laramie Basin of Wyoming - Unavailable since 1909, this publication was originally compiled by the United States Department of Interior. Also included are insights into the mineralization and other characteristics of this important mining region, especially in regards to coal, limestone, gypsum, bentonite clay, cement, sand, clay and copper. **8.5" X 11", 104 ppgs, Retail Price: $11.99**

New Mexico Mining Books

The Mogollon Mining District of New Mexico - Unavailable since 1927, this important publication was originally published by the US Department of Interior and has been unavailable for 80 years. Topics include the geology, rock formations and the formation of ore deposits in this important mining area in New Mexico. Of particular focus is information on the history and production of the ore deposits in this area, their form and structure, vein filling, their paragenesis, origins and ore shoots, as well as oxidation and supergene enrichment. Also included are hard to find details, including the descriptions and locations of numerous gold, silver and other types of mines, including the Eureka, Pacific, South Alpine, Great Western, Enterprise, Buffalo, Mountain View, Floride, Gold Dust, Last Chance, Deadwood, Confidence, Maud S., Deep Down, Little Fanney, Trilby, Johnson, Alberta, Comet, Golden Eagle, Cooney, Queen, the Iron Crown, Eberle, Clifton, Andrew Jackson mine, Mascot and others. **8.5" X 11", 144 ppgs, Retail Price: $12.99**

The Percha Mining District of Kingston New Mexico - Unavailable since 1883, this important publication was originally published by the Kingston Tribune and has been unavailable for over one hundred and thirty five years. Having been written during the earliest years of gold and silver mining in the Percha Mining District, unlike other books on the subject, this work offers the unique perspective of having actually been written while the early mining history of this area was still being made. In fact, the work was written so early in the development of this area that many of the notable mines in the Percha District were less than a few years old and were still being operated by their original discoverers with the same enthusiasm as when they were first located. Included are hard to find details on the very earliest gold and silver mines of this important mining district near Kingston in Sierra County, New Mexico. **8.5" X 11", 68 ppgs, Retail Price: $9.99**

East Coast Mining Books

<u>The Gold Fields of the Southern Appalachians</u> - Unavailable since 1895, this important publication was originally published by the US Department of Interior and has been unavailable for nearly 120 years. Topics include the geology, rock formations and the formation of ore deposits in this important mining area of the American South. Of particular focus is information on the history and statistics of the ore deposits in this area, their form and structure and veins. Also included are details on the placer gold deposits of the region. The gold fields of the Georgian Belt, Carolinian Belt and the South Mountain Mining District of North Carolina are all treated in descriptive detail. Included are hard to find details, including the descriptions and locations of numerous gold mines in Georgia, North Carolina and elsewhere in the American South. Also included are details on the gold belts of the British Maritime Provinces and the Green Mountains. **8.5" X 11", 104 ppgs, Retail Price: $9.99**

Gold Rush Tales Series

<u>**Millions in Siskiyou County Gold**</u> - In this first volume of the "Gold Rush Tales" series, leading mining historian and editor Kerby Jackson, introduces us to the story of how millions of dollars worth of gold was discovered in Siskiyou County during the California Gold Rush. Lavishly illustrated with photos from the 19th Century, this hard to find information was first published in 1897 and sheds important light onto the gold rush era in Siskiyou County, California and the experiences of the men who dug for the gold and actually found it. **8.5" X 11", 82 ppgs, Retail Price: $9.99**

<u>**The California Rand in the Days of '49**</u> - In this second volume of the "Gold Rush Tales" series, leading mining historian and editor Kerby Jackson, introduces us to four tales from the California Gold Rush. Lavishly illustrated with photos from the 19th Century, this hard to find information was first published in 1890's and includes the stories of "California's Rand", details about Chinese miners, how one early miner named Baker struck it rich and also the story of Alphonzo Bowers, who invented the first hydraulic gold dredge. **8.5" X 11", 54 ppgs, Retail Price: $9.99**

More Mining Books

<u>**Prospecting and Developing A Small Mine**</u> - Topics covered include the classification of varying ores, how to take a proper ore sample, the proper reduction of ore samples, alluvial sampling, how to understand geology as it is applied to prospecting and mining, prospecting procedures, methods of ore treatment, the application of drilling and blasting in a small mine and other topics that the small scale miner will find of benefit. **8.5" X 11", 112 ppgs, Retail Price: $11.99**

<u>**Timbering For Small Underground Mines**</u> - Topics covered include the selection of caps and posts, the treatment of mine timbers, how to install mine timbers, repairing damaged timbers, use of drift supports, headboards, squeeze sets, ore chute construction, mine cribbing, square set timbering methods, the use of steel and concrete sets and other topics that the small underground miner will find of benefit. This volume also includes twenty eight illustrations depicting the proper construction of mine timbering and support systems that greatly enhance the practical usability of the information contained in this small book. **8.5" X 11", 88 ppgs. Retail Price: $10.99**

<u>**Timbering and Mining**</u> - A classic mining publication on Hard Rock Mining by W.H. Storms. Unavailable since 1909, this rare publication provides an in depth look at American methods of underground mine timbering and mining methods. Topics include the selection and preservation of mine timbers, drifting and drift sets, driving in running ground, structural steel in mine workings, timbering drifts in gravel mines, timbering methods for driving shafts, positioning drill holes in shafts, timbering stations at shafts, drainage, mining large ore bodies by means of open cuts or by the "Glory Hole" system, stoping out ore in flat or low lying veins, use of the "Caving System", stoping in swelling ground, how to stope out large ore bodies, Square Set timbering on the Comstock and its modifications by California miners, the construction of ore chutes, stoping ore bodies by use of the "Block System", how to work dangerous ground, information on the "Delprat System" of stoping without mine timbers, construction and use of headframes and much more. This volume provides a reference into not only practical methods of mining and timbering that may be employed in narrow vein mining by small miners today, but also rare insights into how mines were being worked at the turn of the 19th Century. **8.5" X 11", 288 ppgs. Retail Price: $24.99**

A Study of Ore Deposits For The Practical Miner - Mining historian Kerby Jackson introduces us to a classic mining publication on ore deposits by J.P. Wallace. First published in 1908, it has been unavailable for over a century. Included are important insights into the properties of minerals and their identification, on the occurrence and origin of gold, on gold alloys, insights into gold bearing sulfides such as pyrites and arsenopyrites, on gold bearing vanadium, gold and silver tellurides, lead and mercury tellurides, on silver ores, platinum and iridium, mercury ores, copper ores, lead ores, zinc ores, iron ores, chromium ores, manganese ores, nickel ores, tin ores, tungsten ores and others. Also included are facts regarding rock forming minerals, their composition and occurrences, on igneous, sedimentary, metamorphic and intrusive rocks, as well as how they are geologically disturbed by dikes, flows and faults, as well as the effects of these geologic actions and why they are important to the miner. Written specifically with the common miner and prospector in mind, the book will help to unlock the earth's hidden wealth for you and is written in a simple and concise language that anyone can understand. 8.5" X 11", 366 ppgs. Retail Price: $24.99

Mine Drainage - Unavailable since 1896, this rare publication provides an in depth look at American methods of underground mine drainage and mining pump systems. This volume provides a reference into not only practical methods of mining drainage that may be employed in narrow vein mining by small miners today, but also rare insights into how mines were being worked at the turn of the 19th Century. 8.5" X 11", 218 ppgs. Retail Price: $24.99

Fire Assaying Gold, Silver and Lead Ores - Unavailable since 1907, this important publication was originally published by the Mining and Scientific Press and was designed to introduce miners and prospectors of gold, silver and lead to the art of fire assaying. Topics include the fire assaying of ores and products containing gold, silver and lead; the sampling and preparation of ore for an assay; care of the assay office, assay furnaces; crucibles and scorifiers; assay balances; metallic ores; scorification assays; cupelling; parting' crucible assays, the roasting of ores and more. This classic provides a time honored method of assaying put forward in a clear, concise and easy to understand language that will make it a benefit to even beginners. 8.5" X 11", 96 ppgs. Retail Price: $11.99

Methods of Mine Timbering - Originally published in 1896, this important publication on mining engineering has not been available for nearly a century. Included are rare insights into historical methods of timbering structural support that were used in underground metal mines during the California that still have a practical application for the small scale hardrock miner of today. 8.5" X 11", 94 ppgs. Retail Price: $10.99

The Enrichment of Copper Sulfide Ores - First published in 1913, it has been unavailable for over a century. Topics include the definition and types of ore enrichment, the oxidation of copper ores, the precipitation of metallic sulfides. Also included are the results of dozens of lab experiments pertaining to the enrichment of sulfide ores that will be of interest to the practical hard rock mine operator in his efforts to release the metallic bounty from his mine's ore. 8.5" X 11", 92 ppgs. Retail Price: $9.99

A Study of Magmatic Sulfide Ores - Unavailable since 1914, this rare publication provides an in depth look at magmatic sulfide ores. Some of the topics included are the definition and classification of magmatic ores, descriptions of some magmatic sulfide ore deposits known at the time of publication including copper and nickel bearing pyrrhitic ore bodies, chalcopyrite-bornite deposits, pyritic deposits, magnetite-ileminite deposits, chromite deposits and magmatic iron ore deposits. Also included are details on how to recognize these types of ore deposits while prospecting for valuable hardrock minerals. 8.5" X 11", 138 ppgs. Retail Price: $11.99

The Cyanide Process of Gold Recovery - Unavailable since 1894 and released under the name "The Cyanide Process: Its Practical Application and Economical Results", this rare publication provides an in depth look at the early use of cyanide leaching for gold recovery from hardrock mine ores. This volume provides a reference into the early development and use of cyanide leaching to recover gold. 8.5" X 11", 162 ppgs. Retail Price: $14.99

California Gold Milling Practices - Unavailable since 1895 and released under the name "California Gold Practices", this rare publication provides an in depth look at early methods of milling used to reduce gold ores in California during the late 19th century. This volume provides a reference into the early development and use of milling equipment during the earliest years of the California Gold Rush up to the age of the Industrial Revolution. Much of the information still applies today and will be of use to small scale miners engaging in hardrock mining. 8.5" X 11", 104 ppgs. Retail Price: $10.99

Leaching Gold and Silver Ores With The Plattner and Kiss Processes - Mining historian Kerby Jackson introduces us to a classic mining publication on the evaluation and examination of mines and prospects by C.H. Aaron. First published in 1881, it has been unavailable for over a century and sheds important light on the leaching of gold and silver ores with the Plattner and Kiss processes. 8.5" X 11", 204 ppgs. Retail Price: $15.99

The Metallurgy of Lead and the Desilverization of Base Bullion - First published in 1896, it has been unavailable for over a century and sheds important light on the the recovery of silver from lead based ores. Some of the topics include the properties of lead and some of its compounds, lead ores such as galenite, anglesite, cerussite and others, the distribution of lead ores throughout the United States and the sampling and assaying of lead ores. Also covered is the metallurgical treatment of lead ores, as well as the desilverization of lead by the Pattinson Process and the Parkes Process. Hofman's text has long been considered one of the most important early works on the recovery of silver from lead based ores. **8.5" X 11", 452 ppgs. Retail Price: $29.99**

Ore Sampling For Small Scale Miners - First published in 1916, it has been unavailable for over a century and sheds important light on historic methods of ore sampling in hardrock mines. Topics include how to take correct ore samples and the conditions that affect sampling, such as their subdivision and uniformity. Particular detail is given to methods of hand sampling ore bodies by grab sample, pipe sample and coning, as well as sampling by mechanical methods. Also given are insights into the screening, drying and grinding processes to achieve the most consistent sample results and much more. **8.5" X 11", 124 ppgs. Retail Price: $12.99**

The Extraction of Silver, Copper and Tin from Ores - First published in 1896, it has been unavailable for over a century and sheds important light on how historic miners recovered silver, copper and tin from their mining operations. The book is split into three sections, including a discussion on the Lixiviation of Silver Ores, the mining and treatment of copper ores as practiced at Tharsis, Spain and the smelting of tin as it was practiced by metallurgists at Pulo Brani, Singapore. Also included is an overview and analysis of these historic metal recovery methods that will be of benefit to those interested in the extraction of silver, copper and tin from small mines. **8.5" X 11", 118 ppgs. Retail Price: $14.99**

The Roasting of Gold and Silver Ores - First published in 1880, it has been unavailable for over a century and sheds important light on how historic miners recovered gold and silver rom their mining operations. Topics include details on the most important silver and free milling gold ores, methods of desulphurization of ores, methods of deoxidation, the chlorination of ores, methods and details on roasting gold and silver ores, notes on furnaces and more. Also included are details on numerous methods of gold and silver recovery, including the Ottokar Hofman's Process, the Patera Process, Kiss Process, Augustin Process, Ziervogel Process and others. **8.5" X 11", 178 ppgs. Retail Price: $19.99**

The Examination of Mines and Prospects - First published in 1912, it has been unavailable for over a century and sheds important light on how to examine and evaluate hardrock mines, prospects and lode mining claims. Sections include Mining Examinations, Structural Geology, Structural Features of Ore Deposits, Primary Ores and their Distribution, Types of Primary Ore Deposits, Primary Ore Shoots, The Primary Alteration of Wall Rocks, Alterations by Surface Agencies, Residual Ores and their Distribution, Secondary Ores and Ore Shoots and Vein Outcrops. This hard to find information is a must for those who are interested in owning a mine or who already own a lode mining claim and wish to succeed at quartz mining. **8.5" X 11", 250 ppgs. Retail Price: $19.99**

Garnets: Their Mining, Milling and Utilization - First published in 1925, it has been unavailable since those days and sheds important light on the mining, milling and utilization of garnets. Included are details on the characteristics of garnets, where they are found and how they were mined. 78 ppgs, 10.99

Gemstones and Precious Stones of North America - Leading mining historian Kerby Jackson introduces us to a classic mining publication on the gems and precious stones of the United States, Canada and mexico. First published in 1890, it has been unavailable since those days and sheds important light on the gems and precious stones that may be found in North America. Included are chapters on diamonds, corundum, sapphire, ruby, topaz, emerald, disapore, spinel, turquoise, tourmaline, garnets, beyrl, peridot, zircon, quartz crystals, feldspars, pearls and many others. Included are details on where these gems and precious stones may be found throughout North America, as well as their characteristics. 360 ppgs, 24.99

Mining Camps and Mining Districts - First released in 1885 by Charles Howard Shinn under the title "Mining Camps: A Study in American Frontier Government", this publication offers a unique look at how early gold miners established their own forms of representative government during the California Gold Rush. Drawing on the the early mining codes of midevial German miners in the Harz Mountains, on the mining customs of the Cornish tin miners and early Spanish mining laws introduced into California, the miners established the first governments in the American West. 340 ppgs, 24.99

BLM Field Handbook for Mineral Examiners - Leading mining historian Kerby Jackson introduces us to a classic mining publication on mine evaluation. First published in 1962, this work sheds important light on the techniques of BLM Mineral Examiners to perform validity on mining claims. 132 ppgs, 10.99

Six Months In The Gold Mines During The California Gold Rush - Unavailable since 1850, this important work is a first hand account of one "49'ers" personal experience during the great California Gold Rush, shedding important light on one of the most exciting periods in the history of not only California, but also the world. Compiled from journals written between 1847 and 1849 by E. Gould Buffum, a native of New York, "Six Months In The Gold Mines During The California Gold Rush" offers a rare look into the day to day lives of the people who came to California to work in her gold mines when the state was still a great frontier. **8.5" X 11", 290 ppgs. Retail Price: $19.99**

The Discovery of Gold in Australia - First published in 1852, it has been unavailable since those days and sheds important light on Australia's gold mining history. Included are rare communications between British agents and the British Crown when gold was first discovered in Australia in 1851. This rare text contains hard to find details on Australia's first mining camps and Britain's early attempts to provide for the orderly regulation of gold mines in that part of the world. Also of interest are hard to find extracts of articles that appeared in the early colonial newspapers that did their best to report on Australia's gold rush as it took place.
102 ppgs, 10.99

www.ingramcontent.com/pod-product-compliance
Lightning Source LLC
Chambersburg PA
CBHW080812180526
45168CB00006B/2422